Praise for *Soil Science for Gardeners*

Robert Pavlis once again proves science is wonderful! *Soil Science for Gardeners* explains why things work in your garden! Equally important, Pavlis uses science to destroy so many gardening myths. This is a book for all gardeners, especially the curious ones.

— Jeff Lowenfels, author, *DIY Autoflowering Cannabis*
and the *Teaming With* Trilogy

Robert Pavlis speaks the language of plants. More to the point, while he understands the science of healthy plant life he knows that none of this knowledge is very useful to the average gardener (nor of much interest) until it is distilled into language that we understand. Voila! *Soil Science for Gardeners*. This book will help readers understand the language of horticulture as it relates to soil. And good soil health is the foundation of every great garden and every successful garden experience. You will become a more confident gardener and your enjoyment of the gardening experience will be that much deeper. I will place my copy of this book within easy reach for future reference. Its pages will become well-worn from use long before it becomes compost. And in my books, composting a good book after it has served its purpose is a high compliment. From book to earth, from earth to life. From life to book. I predict that your reading experience of this book will enlighten and inform you like no other.

— Mark Cullen, C.M., president, Mark's Choice, co-author, *Escape to Reality*

Like many gardeners, I use soil all the time, yet rarely give it much of a thought. But *Soil Science for Gardeners* takes you behind the scenes to the underground world of soil: what it is, its inhabitants, and how to use soil to really get the gardening results you want. Much of what you thought you knew about soil is simply not true and you'll learn why, plus how to make the soil you garden as healthy as possible. Painless science, practical knowledge. Read it from beginning to end!

— Larry Hodgson, the Laidback Gardener

Good gardens start from the ground up. In *Soil Science for Gardeners*, Robert Pavlis tells what's going on in soils and then offers practical directions to manage soil for more productive and healthier plants.

— Lee Reich, author, *The Ever Curious Gardener*

Natural systems are more than the sum of their parts. And yet to understand them we must closely examine these parts. *Soil Science for Gardeners* takes us on a journey through the microscopic levels of soil geology, chemistry, and biology, and illuminates the soil ecology that sustains all life on land. From this knowledge base it then teaches us strategies and methods to regenerate and sustain life giving soil.

— Darrell E. Frey, author, *Bioshelter Market Garden*
and co-author, *The Food Forest Handbook*

Thank goodness for this book! If I had one wish for all gardeners, no matter their experience, I'd wish for them to endeavor to understand the vital role that soil plays in gardening success, and then, I'd have them read this book for the most comprehensive understanding behind it all. As a self-proclaimed soil super-fan, this is the most comprehensive book on the subject I've ever seen. I've been following Robert's work for a long time and I trust his science-based, no-nonsense approach to helping us all debunk gardening myths while making us all smarter gardeners in the process. Robert is one of my trusted, go-to sources when I want to get to the bottom line, and this book does that in spades. *Soil Science for Gardeners* will teach us everything we want and need to know about the critical role soil plays in the success of our garden and landscape, and more. It's extremely well thought out, comprehensive and a joy to read: practical wisdom for real life application in all our gardening pursuits. Three cheers for *Soil Science for Gardeners!*

— Joe Lamp'l, Creator and Host of Emmy Award-winning
PBS television series, Growing a Greener World®
and Founder, joegardener.com and The joe gardener® Show podcast

soil
science
for gardeners

**Working with Nature
to Build Soil Health**

ROBERT PAVLIS

new society
PUBLISHERS

Cover design by Diane McIntosh.
Cover image: ©iStock

Printed in Canada. Third printing, July 2021.

Inquiries regarding requests to reprint all or part of *Soil Science for Gardeners*
should be addressed to New Society Publishers at the address below.
To order directly from the publishers, please call toll-free (North America)
1-800-567-6772, or order online at www.newsociety.com

Any other inquiries can be directed by mail to:

New Society Publishers
P.O. Box 189, Gabriola Island, BC V0R 1X0, Canada
(250) 247-9737

LIBRARY AND ARCHIVES CANADA CATALOGUING IN PUBLICATION

Title: Soil science for gardeners : working with nature
to build soil health / Robert Pavlis.

Names: Pavlis, Robert, author.

Description: Includes index.

Identifiers: Canadiana (print) 20190207140 | Canadiana (ebook) 20190207159
| ISBN 9780865719309 (softcover) | ISBN 9781550927238 (PDF)
| ISBN 9781771423199 (EPUB)

Subjects: LCSH: Garden soils. | LCSH: Soil science.
| LCSH: Soil management. | LCSH: Gardening.

Classification: LCC S596.75 .P38 2020 | DDC 631.4b

New Society Publishers' mission is to publish books that contribute
in fundamental ways to building an ecologically sustainable and just society,
and to do so with the least possible impact on the environment,
in a manner that models this vision.

This book is dedicated to my lovely wife, Judy,
who has always been at my side.

And to my kids, Leah and Mark,
who unfortunately never inherited the gardening gene.
I forgive both of you for not reading this book.

I'd also like to dedicate this book to my English teachers,
who tried so hard to teach me something
but kept failing me in spelling and grammar.
Who knew that, one day, spellchecker would make it all so easy
and allow me to write a book.

Contents

Section 2: Solving Soil Problems

Section 3: A Personalized Plan for Healthy Soil

Introduction

The most common question gardeners ask is What is wrong with my plant?

If a plant is not growing well, there are four potential problem areas. Too little or too much light. That is an easy fix, and new gardeners soon learn to plant in the right light conditions. Watering is the next thing to consider. New gardeners struggle with knowing the correct amount of water to give a plant, but they soon learn that you water when the soil starts to dry out and that you water deep and less frequently.

Pest and disease problems are not simple to diagnose, but you can see most pests, or at least you can see their effect on the plant. Even diseases show symptoms that help you solve the problem. Prevention may not be as easy, but gardeners do learn about common pests and diseases over time.

The fourth area to consider is the soil, and even for experienced gardeners, this remains a mystery. You probably have a vague understanding about nutrients, and you have almost certainly fertilized plants before. But for most gardeners, the stuff that happens underground is a complete unknown.

Go out into your garden and have a close look at it. What do you see? The plants are obvious, especially if they are blooming, but look past the plants at the soil underneath them. You probably don't see anything except soil. You might see some mulch, or a few small stones, but except for the plant, it all looks lifeless.

Pick up a pinch of soil and hold it in the palm of your hand. You can't see them, but you are looking at billions of living organisms representing many thousands of different species. All those life-forms are trying to eke out a living. They are growing, breathing, reproducing, and the larger ones are eating smaller ones, which leads to a lot of excrement that feeds the plants.

Some are trying to attack your plants, while others are forming partnerships with them and defending them from pest organisms. Some are actually roaming the soil, collecting nutrients and delivering them back to plant roots. There are organisms in soil that you can't see, which are spoon-feeding your plants.

One of the reasons soil is so mysterious to gardeners is that our eyes can't see any of this. The number of organisms is so vast and the microbe societies are so complex, we can't get our head around it all. One of the goals of this book is to simplify the soil story and present it in such a way that you truly understand what is going on.

A few years ago, I was designing an introductory gardening course for the general public. I had a look at some large gardening books to get an idea of the main topics that should be covered. One book, of 640 pages, had 4 pages dedicated to soil. Another with over 700 pages did not have a single page on the topic. I decided to start the program by discussing soil and dedicated one-sixth of the course to it.

After 45 years of gardening experience, I realize that growing plants is very easy if you understand the soil below them. It anchors them; it feeds them; and it provides the air and water they need to survive. If you create healthy soil, you can grow anything that is suitable for your climate.

New gardeners, and even more experienced ones, tend to learn about gardening by memorizing rules. When do you transplant a peony? Should you cut back an iris? When is the best time to prune a lilac? These are all rules, and once you learn them, they are easy to follow. Move peonies in fall; cut back German bearded iris in mid to late summer; and prune lilacs after flowering. But there are thousands of different kinds of plants. You will never learn and remember all the rules for all these plants.

A much better approach is to learn the underlying science. Learn how plants grow and the role soil plays. Once you understand that, you can skip learning the rules because you don't need them, and you will be able to grow just about anything. And that is the second goal of this book: I want you to understand what is really going on in soil and how it affects plants. This book paints a simple, clear picture of the natural processes below your feet.

Once you have a really good understanding of the basics, you will be able to evaluate any gardening procedure and determine if it makes sense. For example, once you understand aggregation, you can decide for yourself if tilling is a good practice and if and when it should be used.

More importantly, you will be able to evaluate many of the fad techniques and products that are invented every year. Many of these are simply a waste of time and do not improve soil health or plant growth. You will be a more informed consumer.

What Is Soil Health?

The term is often used, but what does *soil health* really mean? Depending on your interest, it can mean many things. A climate scientist might define healthy soil as one in which the sequestered carbon is increasing. A farmer might define it as soil that produces a good yield. A microbiologist may be measuring microbe populations and diversity.

Gardeners look at plant health. If a plant is growing well, flowering profusely, and has no diseases, the soil must be healthy or at least healthy enough to grow the plant. Some grow well in nutritious soil, while others grow much better in lean sandy soil. The definition of soil health depends very much on the type of plant you are growing.

I am not going to provide a specific definition, but for the purpose of this book, healthy soil is one that grows a wide range of plants, has good aggregation, and supports a high number of microbes. Admittedly, that is a squishy definition, but it is good enough for our purposes.

Using the Book

The book has been divided into three sections, and it is important to follow them from front to back. Section 1, Understanding Soil, provides the basic science background that you need to understand your soil and the interactions between soil life and plants.

This base knowledge is then applied in section 2 that identifies soil issues and provides solutions for them. This is the hands-on section that shows you how to improve your soil.

Everybody reading this book will have different soil issues. Section 3 provides a system that will let you develop a personalized plan for improving your specific soil.

You might be tempted to jump ahead and get into the practical aspects of building healthy soil described in section 2, but it is useful to understand the underlying science. Without this base, you will find it more difficult to select the right action items and the right solutions for your garden.

Terminology

Soil scientists use well-defined terms, but these are not always used in the same way by the general public, which leads to misunderstandings. One of my challenges is to use the terms in this book such that they are useful to the gardener but still reflect the accuracy of the science. To ensure that we are all on the same page, it is critical that we must first agree on some basic definitions.

Organic

The term *organic*—which is used far too frequently to mean several different things—leads to all kinds of misunderstandings. It has become synonymous with "natural," which results in the misconception that anything organic is good for us, our garden, and the planet. The term is used extensively to describe products so that buyers think those are good choices. In the same vein, organic has also come to represent non-synthetic chemicals. In reality, many natural organic chemicals are more toxic than synthetic ones. Most drugs are synthetic and generally are safe, and yet some natural organic chemicals, such as ricin found in caster beans, are some of the most toxic compounds on Earth. Organic does not mean safe.

Organic also refers to agricultural foodstuffs that are produced "organically." This does not mean they are produced without pesticides, or chemicals; it just means that when chemicals are used, they fall under a strict set of guidelines developed by certified organic organizations. If you follow their rules, your operations are organic, even if some of the approved chemicals are synthetic or toxic. The rules become paramount, and safety is secondary.

To a chemist, the word *organic* means something completely different. An organic chemical is any chemical that contains carbon, with the exception of some salts. All sugars, carbohydrates, and proteins contain carbon and are organic, even if they are human-made. Anything that does not contain carbon, including most plant nutrients, is inorganic. By this definition, most synthetic pesticides are organic.

This book will use the chemist's definition of organic and the term *certified organic* to refer to organic agriculture.

The term *organic soil* is used differently by gardeners and soil scientists. For gardeners, it refers to soil that has been treated organically following certain organic certification rules. To soil scientists, it is soil that was created by the layering of plant material instead of the degradation of rocks. It usually contains more than 20% organic matter, and peat bogs and marshes are good examples of this kind of soil. This book uses the latter definition.

Organic Matter

The term *organic matter* is used in a very general sense to refer to any dead flora, fauna, or microbe. This could be recent dead material, such as wood chips and manure, or a highly decomposed form, such as compost or humus.

Fertilizer

The term *fertilizer* can have many definitions. Gardeners often think that the term refers only to synthetic chemical fertilizers, but that is not a correct usage since there are many organic fertilizers that are not synthetic.

Many jurisdictions use a legal definition for fertilizer that requires that the product contain nitrogen, phosphorus, and potassium, and that the amounts of these nutrients are labeled on the package as the NPK value. By this definition, something like Epsom salts would not be a fertilizer even though it provides plants with magnesium. Its NPK value would be 0-0-0, which is not a fertilizer.

In a more general approach, I will use the term *fertilizer* to describe any material that is added to soil with the primary purpose of supplying at least one plant nutrient. I will also use the term *synthetic fertilizer* to refer to human-made chemical products and *organic fertilizer* for natural products.

Fertilizer vs Soil Amendment

A soil amendment is something that is added to soil with the primary purpose to change its physical properties, such as water retention, permeability, drainage, and structure, as well as changes to pH. Lime, for example, is usually applied to change pH, but it also adds some nutrients. Since the primary purpose is to modify the soil, it is a soil amendment, not a fertilizer.

In some cases, the differentiation between *fertilizer* and *amendment* is not clear-cut. Compost is used by most people to add nutrients for plants—as fertilizer—but it also improves the physical properties of soil, so it is also an amendment. Sulfur can be an important component in fertilizer, and it is used to change pH.

Another term in common use is *soil conditioner*. Some try to differentiate between amendment and conditioner, but in this book, they are one and the same and will be called a soil amendment. Many amendments can simply be layered on top of soil, instead of incorporating them into the soil, in which case they are referred to as both amendment and mulch.

Microbes

A large part of this book is focused on the life forms in or on soil. These organisms are classified as flora (plants) and fauna (animals), and these terms work well for the larger ones like mice, earthworms, and perennials, but as their sizes decrease, the difference between plant and animal gets muddy. Many small organisms have some characteristics of both. Most bacteria don't have chlorophyll, so they can't make food from sunlight like plants can, but they don't have many animal characteristics either. And then there are the cyanobacteria that do photosynthesize and the fungi that are more plant-like but don't photosynthesize.

The difference between these organisms is a fascinating subject, but for a gardener, we can keep things simple. I'll use the general terms *microbe* or *microorganism* to refer to this varied group of small organisms.

Understanding Soil

CHAPTER 1

Soil Basics

Nobody has a problem recognizing soil, but it is actually difficult to define. Over the years, its definition has also changed, with the latest one being developed by the Soil Science Academy of America in 2016: "Soil is the top layer of the Earth's surface that generally consists of loose rock and mineral particles mixed with dead organic matter."

It is important to note that this definition does not include the many different organisms that live in soil. This will probably come as a surprise to you since so much is written about the *living soil* and the need to *feed soil*, terms that have led to various misunderstandings about soil.

People make the claim that "soil is alive," and then go on to describe how to feed the soil in order to maintain a balanced health, describing various food that soil wants to eat. The whole idea that soil is a living organism that requires similar attention to animals is completely false and leads to many poor recommendations for managing soil.

Soil is not alive. It does not need to eat or breathe. Soil can be improved to make it better for growing plants, but this is not a health issue in the way animals are either healthy or sick. When people talk about a living soil, they are actually referring to a soil ecosystem that consists of soil and all of the living organisms in and on it. This ecosystem does have life in it, and it supports life, but even it is not alive.

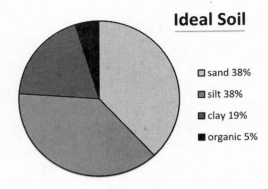

Components of ideal soil.

Components of Soil

If we exclude the small and large rocks, soil has four components: sand, silt, clay, and dead organic matter (OM). The ideal soil contains these in the ratios shown in the diagram. What is a surprise to many is that, by weight, the organic matter makes up only 5%. On a volume basis, it is about 10%. Although OM is only 5%, it is extremely important for providing soil its physical and chemical characteristics.

This concept of ideal soil is a bit misleading since almost nobody has soil with these ratios. You might think it is your job to convert your soil to these ideal ratios, but that is neither practical nor necessary. Consider that almost none of the soil on Earth is ideal, and much of it grows plants just fine. Plants are very adaptable and will grow in most soil.

Changing the amounts of sand, silt, and clay to any great degree is almost impossible because they make up such a large percent of the soil, but you can change the OM level, and since it is a small part of soil, a change of even a fraction of 1% can have a significant effect. The more your soil ratios deviate from ideal, the more issues you will have with your soil.

Origin of Soil

To better understand soil, it is instructive to understand its genesis. Soil starts as large rock formations. Over time these are broken up into smaller and smaller pieces through the action of physical, chemical, and biological weathering. Wind and rain slowly break small pieces off larger ones. Water gathers in cracks, and when it freezes, the rock splits apart. Falling rain picks up CO_2 from the air, resulting in acidic water with a pH of about 5.5, which slowly dissolves some types of rock like limestone. Moss and lichen grow on rocks, and the chemicals they produce through roots slowly dissolve the rock.

It is a very slow process, but one that is continuously taking place. Sand, silt, and clay are just rocks of varying sizes. Eventually, the rock is broken down even more, into its basic elements, which are the nutrients plants use to grow.

As the particles get smaller, it becomes easier for nature to move them around. Rivers and glaciers take rock from one location and move it many miles away. Wind and rain also play a big role in moving soil, especially on sloped ground. The soil at the top of a hill can be quite different from the soil at the bottom. It is estimated that 95% of the Earth's soil has been moved from the area in which it was created.

The soil in your area is the sum of all of these actions. In theory, your soil is made from the bedrock that exists where you live, but it could have moved there from many miles away. Nevertheless, it is still a product of the parent rock that made it.

Granite, sandstone, and shale result in acidic soil because those rocks are acidic. Limestone and basalt, which contain a lot of calcium and magnesium, will result in alkaline soil.

Soil Particles

Soil consists of varying sizes of small rocks, and in order to quantitate them, scientists use the terms sand, silt, and clay. Large pieces, above 2 mm, are called rocks. Sand refers to pieces between 2 mm and 0.5 mm. Silt varies from 0.5 mm to 0.002 mm, and clay is smaller than that. To put those numbers in perspective, if a piece of sand was the size of this page, silt would have the width of one letter and a piece of clay would be smaller than the period that ends this sentence.

Physical Characteristics of Sand, Silt, and Clay.

Property	Sand	Silt	Clay
Particle size (mm)	2.0–0.05	0.05–0.002	<0.002
Visible to naked eye	Yes	No	No
Cohesion (attraction to each other)	Low	Moderate	High
Ability to hold water	Low	Moderate	High
Rate of water infiltration	Rapid	Slow	Slow
Degree of aeration	Good	Moderate	Low
Resistance to pH change	Low	Moderate	High
Ability to hold nutrients	Low	Low	High
Compactability	Low	Moderate	High

Sand

Sand is large enough to be seen with the naked eye, and if you rub some soil between your fingers, you can even feel its gritty nature. Sand particles can exist in many shapes, but you can think of them as being round.

The physical property of sand is a function of its size and shape. It does not pack very well, and it has a lot of air space between the particles. When you walk on sand, it tends to move out of the way rather than compact. Sandy soil is easy to dig and is usually described as light and crumbly.

From a chemical perspective, the surface of sand is very stable. It does not react with most chemicals, and water doesn't stick to it very well, which is one of the reasons it drains quickly.

Silt

Silt is too small to be seen by eye, and if you rub it between your fingers, it feels very smooth, like talcum powder. You can think of silt as being small sand particles. They are chemically stable, but due to the smaller size and smaller air spaces between them, water does not drain as quickly as in sand. Silt also compacts more easily than sand.

The properties of silt tend to be halfway between those of sand and clay. Like sand, its particles do not stick to one another, and they don't hold water tightly.

Clay

Clay consists of the very smallest particles, including anything below 0.002 mm. They are so small they can't even be seen with a light microscope.

Whereas sand and silt can be thought of as round particles, clay consists of flat thin plates that pack very close together. Think of a loaf of sliced bread with each slice being a clay particle. These particles are highly charged, which makes them react with themselves and other chemicals. Water and nutrients stick to them.

When you handle clay, it sticks together, and you can easily role it into a ball that retains its shape. This makes it fun to play in but not so great when you have to dig in it. A shovel of clay soil tends to retain its shape, forming clods of soil.

As a result of its size and chemical properties, clay behaves quite differently from sand and silt. It holds water very tightly, and even though the spaces between the particles are small, the total volume of space is large, allowing clay to hold significant amounts of water. This, along with its charged nature, means that clay takes a long time to drain.

Clay also expands when it gets wet and shrinks when it dries. You can see this in the form of cracks on the surface of the soil. Clay is easily compressed, especially when it's wet. Walking on clay squishes the slices of bread even closer together, and once compressed, it is difficult to decompress them by physical means.

Unlike sand and silt, which result from rocks breaking up into smaller and smaller particles, clay is formed when very small minerals stick together to form larger particles. Since clay is created, it has the properties of the parent minerals, and there are many different kinds of clay. As clay is formed, it also incorporates particles of organic matter and plant nutrients.

Soil Texture

We have talked about sand, silt, and clay as if they occur as isolated soils. This can happen in special situations, but it is much more likely that they occur together. Such soil will have properties somewhere between the extremes.

Scientists use the term *soil texture* to describe the ratio of sand, silt, and clay in a given sample, as illustrated in the soil texture diagram.

Ideal soil, which has 40% sand, 40% silt, and 20% clay, is referred to as loam. If the soil contains a larger amount of clay, it is called clay loam. A very sandy soil with some clay and silt is called sandy loam.

These soil designations are helpful when referring to a particular sample, or when you have a discussion with someone else with different soil because a sandy loam will function differently than a clay loam. Knowing your texture, which is easily determined by the soil texture test, will help you understand the physical and chemical properties of your soil.

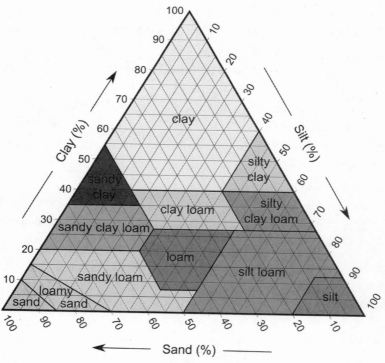

Soil texture triangle.
Credit: https://commons.wikimedia.org/wiki/File:USDA_Soil_Texture.svg

Importance of Particle Size

Particle size affects two important soil characteristics: the surface area and the amount of pore space (both number and size of spaces between particles). Large particles have a relatively small surface area with few large pores. Small particles have a large surface area and many small pores. These properties affect the way water travels through the soil, and it impacts where and how roots grow.

After a rain, most of the pore spaces are full of water, but that soon runs away. What is left is a lot of soil particles that are each covered in a thin film of water that lasts for days, and even weeks. It is this thin layer of water that plant roots use to get their moisture.

Sand has a relatively small surface area, compared to clay. To put this into perspective, a handful of sand has a surface area equivalent to the top of a table, whereas a handful of clay has a surface area equal to a football field. For this reason, clay holds much more water and stays wet much longer, making it easy for roots to get water for a longer time.

Nutrients also stick to the surface of particles, so a larger surface area also means that there are more nutrients available to plants. This is one reason why clay is very nutritious.

Air and Water

The term *soil solution* is used to describe the water that is attached to the surface of soil particles.

Air and water are critical for proper plant growth. What most people don't realize is the large amount of both in soil. The actual amount depends on several factors, such as its texture, the amount of organic material, and the degree of compaction, but ideal soil contains about 25% air and 25% water.

Immediately after a heavy rain, much of the air is forced out and replaced with water. Gravity, evaporation, and plants will reduce the level of water, which is then replaced with air. Perfectly dry soil will have no water and 50% air. Such dry soil is rare and is mostly found in laboratories. Most soil holds some water, even if plants are no longer able to get any of it.

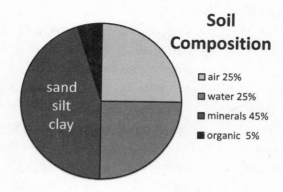

Soil Composition

- ☐ air 25%
- ▨ water 25%
- ▩ minerals 45%
- ■ organic 5%

sand
silt
clay

Composition of ideal soil.

Evaporation is the process by which liquid water turns into water vapor and escapes into the air. This happens right at the surface of the soil, which is why the top layer of soil can be quite dry while the soil is still quite wet a few inches down. As evaporation takes place, more water will be drawn to the surface by capillary action. This process slowly dries out the soil.

Gravity is also at work, pulling water down deeper into the soil. Eventually it is down far enough to enter reservoirs deep in the ground, or, depending on topography, it might flow into a river or lake.

Plant roots are constantly absorbing water and transferring it to their leaves, where much of it evaporates through leaf openings called *stomata*. Some is also used in chemical processes like photo-synthesis. This effect of plants is significant. A large tree can remove up to 100 gallons (400 liters) of water a day and discharge this into the air as water vapor.

As water leaves the soil, the pore spaces are filled with air. As soil dries, more and more air enters the soil. This cycle is repeated next time it rains or you irrigate. It is important to understand that there is always some air remaining in soil, and this is critical for plant health since roots need oxygen.

Most people have heard about photosynthesis, a process in which plants absorb carbon dioxide (CO_2) from the air, produce sugars using sunlight, and then give off oxygen (O_2). This is the opposite of what animals do. They absorb O_2 and give off CO_2 in a process called respiration.

What you may not realize is that plants also respire, absorbing O_2 and giving off CO_2.This happens not only at night but also during the day and for the same reason animals do it. The process allows plants to convert sugars into energy, which they need for growth. This takes place in all their parts, but a lot of it happens in the roots; they need to be able to absorb oxygen from the soil, or they die.

This explains why many plants can't grow in areas that are constantly wet. This kind of soil does not provide enough oxygen for roots. It also explains why plants die if you water too much and why some plants just don't grow well in clay soil that does not hold enough air.

Chemical Nature of Water

The molecular formula for water is H_2O: two hydrogen atoms for every one of oxygen. But the three-dimensional structure of water is much more interesting than this: it looks like a wide V, with the oxygen at one end and the two hydrogen atoms at the other. Because of this structure, one end of the molecule has a positive charge and the other has a negative charge. You can think of water molecules as being small magnets, with one end being attracted to the other. This sticky nature of water, called cohesion, plays a key role in understanding how water and nutrients move through soil.

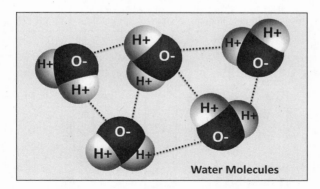

Water molecules react like little magnets.

Movement of Water Through Soil

It has just rained, and the top layer of the soil is saturated with water. Why does this water not run away? Why does gravity not pull it straight down?

The answer lies in the cohesive properties of water. The electrical charges on the water molecules are much stronger than the force of gravity. In effect, the water molecules hold on to each other and to soil particles so tightly that gravity can't pull them down very far.

Remember the evaporating water at the surface of the soil? As the surface water molecules evaporate, they pull the next lower molecules higher because of cohesion. And since each of these are connected to even more molecules lower down, many are pulled up, preventing gravity from pulling the water lower.

There is also another force at play. Clay and organic matter also have negative charges. The positive ends of water molecules stick to the negative charges on clay and OM. These in turn hold onto other water molecules, making it even harder for gravity to pull the water down.

It is interesting to watch what happens when water is dropped onto soil: it does not run straight down as you would expect. The water does moves down, but at the same time, it moves sideways in all directions due to the interplay of charges on clay, OM, and water.

Sand particles have large pore spaces and almost no charge. Water will form a very thin film around each sand kernel, but since sand is not charged, this film is not held very strong.

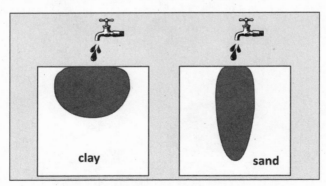

Movement of water in clay and sand.

Water can hold onto itself, but there is a limit in its ability to do this. The pore spaces between particles of sand are just too large for water to fill the space and stay there. Gravity takes over and pulls water out of these spaces fairly quickly. This explains why sand drains so fast. When water is dripped onto sand, most of it runs down, but a bit also runs sideways due to the cohesive nature of water.

Clay soil behaves quite differently. Clay particles are negatively charged, and the pore spaces are very small. Water sticks to clay very tightly and completely fills the pore spaces. Gravity is no match for these forces, and as a result, water dripping onto clay moves sideways much more than in sand. These properties of clay also reduce the rate of evaporation.

So far, we have been discussing pure sand and clay soil, but real soil also contains organic matter that also has negative charges, just like clay. Water also sticks to OM very tightly. As an example, silt loam with 4% OM has twice as much water available as the same soil with 1% OM, illustrating the important of organic matter for plants.

Over time, evaporation, gravity, and plants do reduce the level of water in soil. The large pores are emptied first, then the medium-sized pores, and finally the very small pores. Plant roots are relatively large and can grow only in the large and medium pores. They just don't fit in the small pores.

What happens when a plant root grows into a pore space? The surface of the root comes in contact with the water layer around the soil, and it is able to absorb the water. Remember those little water magnets and how they reacted with evaporation; the same happens here. As a water molecule is pulled into the root, it pulls more water into the space that was vacated. This results in a constant flow of water, from areas away from the root toward the root.

It is the cohesive nature of water that allows roots to pull water from small pores, even though they are too big to fit into the spaces. This all works great, but in time, the water layer gets thinner and thinner because there is just less water in the soil. At some point, the flow of water stops, and roots just can't get enough water.

Plants grow new roots where they are able to get water, nutrients, and oxygen. As the top of the soil dries, new growth happens deeper in the soil where they can still access water. Watering deeply and less frequently results in the development of deep roots that allow the plants to survive dry periods.

Aggregation and Soil Structure

So far we have looked at soil on a microscopic level, but soil is much more than that. If you go into an established woodlot and get your hands in the soil, you will notice that it does not really feel like a bunch of sand, silt, and clay. It consists of large crumbly pieces that are dark in color. It is very airy with lots of pore spaces of all different sizes.

What you are looking at is the macro structure of good soil. The smaller pieces of sand, silt, and clay have been mixed with OM to form larger structures called *aggregates*. Aggregation is not well understood by gardeners, but it is a very critical part of good soil. When soil has it, you will grow lots of plants. When aggregation is lacking, the soil performs poorly. Creating good soil is all about improving aggregation.

The key to aggregation is a special sauce that goes by various names, including binding agents, mucus, organic binders, and organic cement. This binding agent acts like glue to stick the sand, silt, clay, and organic matter together into large soil particles, which soil scientists call *peds*.

The binding agent consists of many different kinds of chemicals produced by living organisms. You can think of them as life juices. Plants, bacteria, fungi, earthworms, and small insects all excrete juices, and some of the chemicals work great as a glue.

Clay, iron oxide, and OM also act as cement in forming aggregates, but microorganisms provide the best binding agents. Sodium reverses the aggregation process and can even prevent it from forming in the first place. This is one reason that irrigation with sea water is a problem.

Aggregation is a two-step process. First, binding agents from various microbes help stick very small particles together to form microaggregates. Then the microaggregates are stuck together into macroaggregates by the mycelium of fungi and hyphae of actinomycetes. There is a direct relationship between size of aggregates and the fungal biomass. The cultivation of soil breaks down fungi mycelium, resulting in a marked decrease in macroaggregation.

Why are aggregates so significant to gardeners? Aggregates indicate a healthy soil system since they form only if the amount of dead organic matter or living microbes is high enough, both being critical for plant growth. More importantly, large pore spaces that are a perfect size for roots are created between the aggregates. The larger pore spaces also make it easier for water and air to move through the soil, increasing drainage and providing oxygen for plant roots.

The structure of aggregates is fairly loose and contains many small pores that hold a lot of water. Microbes also use these smaller channels to hide from larger organisms.

Earlier in this section, I described the soil from a virgin forest. I strongly suggest you visit one and play in the soil. You will be able to both see and feel the aggregates. Look at the various sizes of particles. Feel how crumbly they are, and how easily you can move your

Soil Test: Degree of Aggregation

Remove surface debris from the soil. Try to insert your fingers into the soil. If this can be done easily, you have either a very sandy soil or a high level of aggregation.

Dig up a shovelful of fairly dry soil. Can you see large crumbly pieces, or does it look like sand and silt? Compare the look to the soil in an established wooded area. If you gently squeeze one of the larger pieces, does it fall apart easily? You want to see large pieces of aggregation that can be easily broken apart into smaller pieces that are still larger than sand.

Sandy soil with no aggregation falls apart into sand, and the particles are clearly not stuck together. Clay soil without aggregation forms clods that are hard to break apart, and you can't see any mid-sized particles.

fingers through them. The long-term goal for your garden is to produce the same kind of soil.

Aggregation is a continuous process that is either improving or getting worse. The natural binding agents slowly decompose and need to be continuously replaced by microbe activity. Provided the rate of activity is high enough, aggregation continues to get better.

Soil could have significant aggregation, but the aggregates might not be very stable, which means they easily break apart at the slightest change in soil condition. Very stable aggregates can take some destructive forces and remain intact. The aggregate stability test, also called Archuleta slake test, can be used to measure quality.

Soil Test: Aggregate Stability

Dig up a shovelful of fairly dry soil. Gently break it apart into large pieces, and find a chunk about the size of a golf ball. Set it on some paper and let it air-dry for at least 48 hours so the test can be done on dry soil.

Prepare a clear glass jar that is at least 5 inches high. Cut a piece of wire mesh with ¼-inch holes, and use it to form a basket that will sit on top of the jar, as show in the picture. Fill the jar with water so the water is several inches above the basket.

Gently place the chunk of soil into the basket and wait. Water will rush into the pores, causing significant shear forces. A stable aggregate will remain intact. A poorly formed aggregate will fall apart.

Wait for 1 minute. If the aggregate has mostly fallen apart, score it a 3 for poor stability. If it breaks down in 1 to 5 minutes, score it a 2. If it is still mostly intact after 5 minutes, it is very stable, and you can score it a 1.

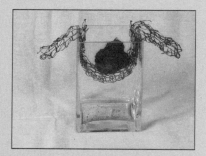

Jar and bracket used in aggregate stability test.

Soil pH

Soil pH can have a significant effect on plant growth, which is why so much has been written about it.

The pH scale measures the number of hydrogen ions in a liquid, given as a number between 0 and 14, with a pH of 7 being neutral. Anything above 7 is alkaline (low level of hydrogen ions), and below 7 is acidic (high level of hydrogen ions). Most plants grow best at a pH between 5.5 and 7.0, but some plants prefer values above or below this. Soil pH ranges from 3.5 to 10.5, but these are extremes; most are between 5 and 8.5.

This is fairly common knowledge, but what most people don't know is that pH is measured on a logarithmic scale. (Do you remember high school math?) This means that a pH of 8 is 10 times more alkaline than a pH of 7; a pH change of 1 unit is actually an acidity change of 10 units. A change of 2 numbers, for example 5 to 7, is a change of 100, which is a significant change.

The pH of your soil depends very much on the rock that formed it. Granite, sandstone, and shale are acidic rocks and produce acidic soil. Limestone and basalt are alkaline rocks and produce alkaline soil. They continue to decompose, and as long as some rocks remain in your soil, it will be difficult to change the pH.

Calcareous soils are a bit different. They have an underlying layer of chalk or limestone rock and contain a high level of calcium carbonate ($CaCO_3$). Organic soil is also different because it was created by the layering of plant material instead of the degradation of rocks. It usually contains more than 20% organic matter and is found in peat bogs and marshes.

The pH of local soil is also affected by rain in two ways. As it falls through the sky picking up CO_2, it forms carbonic acid, and by the time this hits the ground, the pH of the rain is about 5.5. If it also picks up pollutants as it drops, it can be even more acidic.

A high rainfall, as exists in the eastern US, keeps soil acidic by washing minerals (cations) deeper into the soil, which increases the relative number of hydrogen ions near the surface. In dry regions like the western US, a lack of rain results in water moving from lower levels up to the surface due to evaporation, carrying up minerals like calcium and sodium, resulting in alkaline conditions.

Other natural processes can acidify soil:

- Respiration by soil roots and soil organisms produces CO_2, which is acidic.
- Decay of organic matter produces organic acids.
- Plant growth absorbs minerals from the soil leaving behind hydrogen ions.

The pH of soil can also be changed manually:

- Sulfur decreases pH.
- Lime increases pH.
- Fertilizers can increase or decrease pH.
- Compost and manure can change pH.

The reality is that changing soil pH long-term is much more difficult than claimed in books and on the internet. In most cases, it quickly reverts back to its original pH.

You have probably seen suggestions that the ideal pH range for plants is 6 to 7. This is true for mineral soils (made from rock), but for organic soils (peat and marsh bogs), a better range is 5.5 to 6.

Soil Myth: Soil pH

People talk about "soil pH," but what they are actually referring to is the pH of the water surrounding the soil particles—the soil solution. It is also an average of the total soil area. Specific locations within the soil can be quite different. A spot with organic matter and lots of bacterial activity will have a different pH from an inch away where there is less organic matter. The pH right next to a clay particle can be as much as 1 pH unit different from the water that bathes the clay.

The rhizosphere, the area right next to a plant root, can have a very different pH from the soil solution pH.

pH and Nutrient Availability

Given all of the talk about pH and plant growth, you would think that pH is essential to plant growth, but in reality plants are not directly affected by the level of hydrogen ions. The issue with pH is that it affects the nutrient levels in the soil solution and therefore influences plant growth. The chemistry gets a bit complicated, so I'll just provide a few examples to illustrate this.

Imagine a soil that has pH 7 and plenty of nutrients for plants. If we add some hydrogen ions, the pH will become more acidic. At this lower pH, phosphorus reacts with aluminum and precipitates, reducing the number of both ions in the soil solution. Plants now can't get enough of either one to grow properly. There is still lots of phosphorus and aluminum in the soil, but it is now in a form that plants can't use.

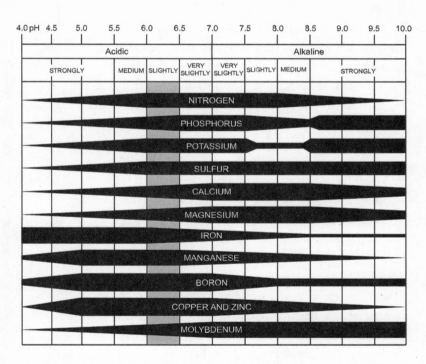

pH affects nutrient availability. Credit: https://commons.wikimedia.org/wiki/File:Soil_pH_effect_on_nutrient_availability.svg

If, however, we make our imaginary soil alkaline, calcium reacts with iron and reduces the amount of iron ions in the soil solution. Plants now show interveinal chlorosis due to a lack of available iron. Iron is essential for making chlorophyll, the green part of leaves, and when plants can't get enough, they develop a yellow coloration in the spaces between the veins. The problem is not a lack of iron but a lack of available iron.

Most plant nutrients are affected by pH. In general, nutrients are more available at a neutral pH and become less available at extreme pH values. The chart shows how the availability of various nutrients changes with varying pH. Keep in mind that it shows relative amounts and can't be used to calculate the amounts available at any given pH.

pH and Toxicity

Plants need certain nutrients, but these same nutrients can become toxic if they exist in concentrations that are too high. At a low pH, aluminum and manganese can reach toxic levels.

Aluminum ions are found in the soil solution and inside clay particles, but the amount in the soil solution at higher pH is normally low. As pH becomes more acidic, the aluminum inside the clay particles is released and enters the surrounding water, dramatically increasing the concentration. At some point, aluminum becomes toxic and inhibits root growth. Aluminum toxicity is not usually a problem in mineral soil that has a low level of clay, or in organic soil.

Aluminum sulfate is a commonly recommended fertilizer, but due to possible toxicity, it should not be used on clay soils.

pH and Soil Organisms

Organisms living in soil are also affected by pH, and most grow best in neutral soil. Bacteria and actinomycetes prefer a slightly alkaline pH, and fungi prefer it a bit more acidic. Earthworms and larger organisms need a pH above 5 and do best near 7.

Specialty microbes can be found at almost any pH. Bacteria exist at both pH 1 and pH 11.

CHAPTER 2

Plant Nutrients

Plants use nutrients from two main categories: non-mineral and mineral (originating from minerals). The non-mineral nutrients make up 96% of a plant and consist of oxygen, hydrogen, and carbon. The plant absorbs CO_2 from the air, which provides most of the carbon and some of the oxygen it needs, and takes in water and oxygen through the roots. The water is transported to leaves, where it is broken down into oxygen and hydrogen. Provided that the soil has a normal amount of water and air, a healthy plant has no problem getting all of the non-minerals it needs.

Although mineral nutrients, like nitrogen, phosphorus, and potassium, account for only 4% of a plant's weight, they are vital to its growth. The rest of this chapter will focus on mineral nutrients.

Plants have four sources for mineral nutrients: soil minerals, organic matter, nutrients adsorbed onto clay and humus, and the soil solution. These nutrients are generally not available to plants directly; organic matter needs to decompose before this occurs. The nutrients adsorbed to soil and humus can be accessed by plants, but it is more likely that these enter the soil solution, the water that surrounds all the particles in soil and the plant roots, before they get to the plant.

Ions

Before we discuss nutrients in more detail, it is important to understand some chemistry. You probably recognize the names of many nutrients, such as iron, zinc, potassium, calcium, and magnesium, but you may not know that these are all metals. For example, pure calcium is a dull silvery-gray and looks like iron.

We talk about plants using iron, calcium, and magnesium, but actually plants can't use any of these. Putting an iron nail in the soil does nothing to feed plants. They are able to use these metals only once they are converted into something called *ions*.

When calcium is exposed to air, a chemical reaction with oxygen forms a type of rust, producing a white powder, calcium oxide (CaO), which contains one molecule each of calcium and oxygen. When CaO dissolves in water, calcium and oxygen separate into charged particles that are called ions. The calcium ion has a positive charge (a *cation*), and the oxygen ion has a negative charge (an *anion*).

Remember the water magnets? One end of the water molecule has a positive charge, and the other has a negative charge, and it is this property that allows the calcium oxide to separate into two charged particles. The calcium, being positive, is attracted to the negatively charged end of the water molecule, and the oxygen is attracted to other.

Once the calcium is in the form of an ion, plants can absorb it through the roots and use it inside the plant. We talk about plants using calcium, but what they really use are calcium ions. This may seem like unimportant semantics, but it is critical for understanding how nutrients behave in soil and how they become available to plants.

All of the mineral nutrients used by plants form some type of ion. Some are a bit more complex than the above calcium example, but the principles are exactly the same. This table shows a list of the macronutrients and their ion forms.

What Is Salt?

The general public uses the term *salt* to mean table salt, which is sodium chloride. Chemists, soil scientists, and this book use the term to refer to any compound that is made up of ions. Sodium chloride is one of many different types of salts. In water, it breaks up into sodium ions ($Na+$) and chloride ions ($Cl-$).

Chemical and Ion Form of Nutrients.

Nutrient name	Chemical symbol	Ion form	Ion name
Carbon	C	none	
Hydrogen	H	H+	
Oxygen	O	none	
Nitrogen	N	NO_3^-, NH_4^+	Nitrate, ammonium
Phosphorus	P	HPO_4^{-2}, $H_2PO_4^-$	Phosphate
Potassium	K	K^+	
Calcium	Ca	Ca^{+2}	
Magnesium	Mg	Mg^{+2}	
Sulfur	S	SO_4^{-2}	Sulfate

Compounds such as ammonium nitrate and potassium phosphate, found in synthetic fertilizer, are also salts, as is calcium oxide. Urea fertilizer is an organic molecule made up of carbon, hydrogen, oxygen, and nitrogen, and since it does not form ions in water, it is not a salt.

Table salt or road salt releases sodium ions in water. All life, including plants, needs some sodium, but as the amount of sodium in soil increases, it can quickly become toxic to plants and microbes. It is best to keep sodium out of the garden.

Salt can also be harmful to plants for a reason that is related to osmotic pressure. Ions in water act like strong magnets and pull water toward themselves. If one area has plenty of ions and a neighboring area has very few, water will move from the area with few ions

Soil Myth: Salt Kills Soil Microbes

Many believe that the salts in synthetic fertilizer harm soil life, but that is not true. They dissolve in water, forming ions, exactly the same as the ions released from organic material like manure or compost. They are essential for the growth of microbes and plants.

Any chemical, no matter how useful, can become toxic at high levels. Even too much water will kill you. Provided fertilizers are used in appropriate amounts, they do not harm soil life.

to the one with more ions because a higher concentration of ions has a stronger pull for water. Plant roots use this phenomenon to their advantage. They keep a high concentration of ions inside the root, and this normally causes water to flow from outside the root into the root, which they then transport into the rest of the plant.

What happens if too much fertilizer is used? The salts in the fertilizer are released as ions that make the concentration of ions outside the roots higher than inside. This causes water to flow from inside the root out into the soil solution. With this loss of water, the plant experiences what is equivalent to draught conditions, and the upper leaves start to dry out. This is why lawn grass goes brown if you use too much fertilizer. The blades of grass look burnt because they can't get enough water from the roots.

Movement of Nutrients in Soil

Sand and silt particles have almost no electrical charge on their surface, so ions don't stick to them very well. When nutrient ions come into contact with these particles, they just keep moving along with the water. For this reason, rain easily washes nutrient ions out of sand and silt into the subsoil layers, which explains why such soils have low natural fertility.

Clay and organic matter have both negative and positive charges, but they are mostly negatively charged. Both anions and cations stick to clay and OM very tightly, preventing water from washing them away. When rain flows through clay soil or soil that contains a lot of OM, it does dislodge some nutrients and moves them deeper in the soil, but the effect is minor. Most nutrients remain stuck in place. The net effect is that nutrients move much more slowly in these soils than in sandy soil.

Essential Plant Nutrients

Essential nutrients are those that plants must absolutely have in order to survive. They include carbon, oxygen, and hydrogen, which are obtained from air and water. The other essential nutrients, including the following, are absorbed through roots.

Soil Myth: Organic Nutrients Are Better

It is believed that nutrients from organic sources are much better for plants than nutrients from synthetic fertilizer. This very common myth that is promoted by the organic movement is completely wrong.

Most synthetic fertilizer consists of salt compounds. Good examples are ammonium nitrate, calcium carbonate, and potassium phosphate. When these dissolve in water, they separate into ions, namely ammonium, nitrate, calcium, carbonate, potassium, and phosphate. Plant roots can absorb all of these.

When an organic source, like manure or compost, is added to soil, it slowly decomposes into the same ions found in synthetic fertilizer. The nitrate ion from an organic source is exactly the same as a nitrate ion from a synthetic fertilizer. Neither labs nor plants can tell the difference between the two sources once the ion has been released into water.

Once you understand this, it becomes clear that both sources result in exactly the same nutrients. Nutrient ions can originate from an organic source, but they can't be any more organic than the ones from fertilizer. They are all inorganic.

Used in larger amounts, macronutrients include nitrogen (N), phosphorus (P), potassium (K), calcium (Ca), sulfur (S), and magnesium (Mg). Because many soils contain plenty of Ca, S, and Mg, the main ingredients in fertilizer are N, P, and K.

The micronutrients (or trace minerals) are boron (B), chlorine (Cl), manganese (Mn), iron (Fe), zinc (Zn), copper (Cu), molybdenum (Mo), nickel (Ni), and cobalt (Co). There is still some debate as to whether silicon (Si), nickel (Ni), chlorine (Cl), and cobalt (Co) are essential.

Plants can use additional nutrients that are nonessential. This means plants will use them if available, but they do not need them in their diet. In some cases, they are found only in specific types

Soil Myth: Plants Use Lots of Different Nutrients

Numerous products on the market, including fish fertilizer, sea salt, and rock dust, promote the idea that more nutrients are better. For example, a product containing Australian sea salt claims that it contains 99 elements that are beneficial for plants. Some claim that since fish live in the ocean, they can also provide all of these nutrients. Azomite, a rock dust, has 67 essential minerals.

According to Stamford University, sea salt contains only 42 elements, or 47 minerals. A plant uses at most 21, and some of these are not essential. Adding nutrients that plants can't use is a waste of resources.

of plants. Nonessential nutrients, also called *beneficial nutrients*, include aluminum (Al), selenium (Se), sodium (Na), vanadium (V), and gallium (Ga).

This brings the total of useful mineral nutrients to 21 (not counting oxygen, hydrogen, and carbon).

Nitrogen

Nitrogen is the key nutrient for you to be concerned about as a gardener. Next to carbon, hydrogen, and oxygen, it is the most abundant nutrient in plants and the one that soil is most likely lacking, especially in cooler climates.

Nitrogen is used as a building block for all kinds of large molecules. Enzymes, which are a special type of protein, contain 16% nitrogen. Since enzymes cause chemical reactions, they control just about everything that happens in a plant, including photosynthesis. Nitrogen is also part of DNA, RNA, and chlorophyll—all vital for plant growth.

Commercial fertilizer can contain a variety of different chemical forms of nitrogen, including ammonium (NH_4), nitrite (NO_2), nitrate (NO_3), and urea (CH_4N_2O), but plants can directly use only nitrate and ammonium. Trees and shrubs tend to use more ammonium, and vegetables and grain crops prefer nitrate.

Why would fertilizer contain nitrite and urea if plants can't use them? The answer involves an important concept in gardening. Microorganisms in the soil are able to convert one form into another, which is happening all the time. This conversion happens so quickly that by the time you get a soil sample to a lab, it would have already changed so much that the analysis would be meaningless. If you want nitrogen analyzed, you need to freeze the sample as soon as you take it and transport it to the lab frozen.

Ammonium, nitrite, nitrate, and urea are all very soluble in water, and once dissolved, they move along with the water. Rain and irrigation easily wash nitrogen out of the soil and into rivers and lakes causing pollution.

What does all this mean for the gardener? Nitrogen is a critical nutrient for your plants, but it is very unstable in soil. It is easily and quickly converted from one form to another and washed away. Your plants can have lots of nitrogen available in the morning and after a good rain have a shortage. As a gardener, you never know how much nitrogen is available to your plants because the amounts change so quickly. Nitrogen is the nutrient that is most likely to be in short supply for your plants.

Although nitrogen is important, too much is not good either because it creates soft, weak growth that is more prone to disease and pests, slows ripening of crops, delays hardening off for winter, and can affect flavor in vegetables. Too much also encourages growth of roots, stems, and leaves instead of flowers and fruit.

So far we have talked about the plant-available forms of nitrogen, but that accounts for only about 3% of the nitrogen in soil. The rest is tied up as large molecules in both living and dead organic matter. Decomposition of organic matter annually produces about 90 pounds of nitrogen per acre (100 kg/hectare).

Nitrate and nitrite are both anions with a negative charge and are quickly leached through soil. Ammonium is a cation, and its positive

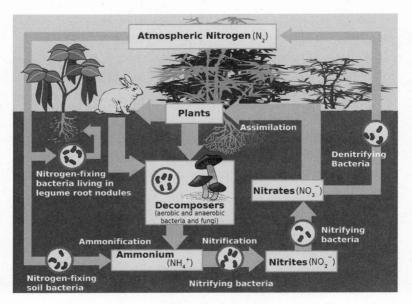

Nitrogen cycle.
Credit: https://commons.wikimedia.org/wiki/File:Nitrogen_Cycle.svg

charge helps it stick to negatively charged clay and OM, slowing down the leaching process. For this reason, ammonium, or urea that is converted to ammonium in soil, is a better choice for fertilizer.

Nitrogen Cycle

Atmospheric nitrogen, N_2, is a gas that makes up 70% of the air we breathe, but most plants and animals can't use it. Certain bacteria, blue-green algae, and actinomycetes are able to convert N_2 into ammonium through a process called *nitrification* or fixation of nitrogen. Other microbes convert the ammonium to nitrite and then to nitrate.

Once nitrogen has been converted into ion forms, plants and microbes immobilize it into large molecules, providing a way to keep it in the soil. When these organisms die, the organic matter slowly decomposes, making the nitrogen available to other organisms. In this way, nitrogen cycles through one organism after the other, a process that is critical to all life on Earth.

Some special organisms, denitrifying bacteria, are able to convert nitrogen ions back into atmospheric nitrogen, which removes the nitrogen from the soil.

Phosphorus

Phosphorus (P) is a critical part of plant growth. It is part of DNA and RNA, the genetic material in plants, and it plays a critical role in cell walls. It is also a component of a molecule called ATP, which is like a rechargeable battery for cells, storing and releasing energy as it's needed. ATP is involved in many reactions taking place, including photosynthesis.

Phosphorus naturally comes from dissolved rock, and with the exception of sandy soil, it is present in most soils. Unlike nitrogen, it is not very soluble in water, and so natural forms of it do not leach out of soil quickly. In fact, most phosphorus sticks to soil so tightly that it is difficult for plant roots to extract it.

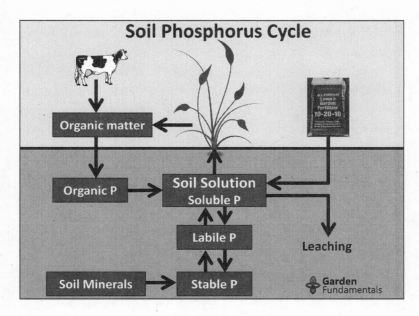

Soil phosphorus cycle.

Two basic forms of phosphorus exist in soil: free phosphorus and phosphorus that is tied up in both dead and living organic matter. The former exists in three different forms in soil:

- Soluble P is dissolved in the soil solution and is available in very small quantities.
- Labile P is held loosely by soil particles.
- Stable P makes up the majority of free phosphorus and is held strongly by soil particles.

Plants can access the soluble and labile P through their root system, but it is easiest for them to use the soluble form. As they use up the soluble P, some of the labile P is converted to soluble P so that there is always some available to plants.

Commercial fertilizer is mostly soluble P in the form of phosphate. When added to soil, this is converted to labile P within 24 hours and eventually becomes stable P. This means that most of the phosphorus fertilizer you add to the soil is available to plants for a short period of time, and then it gets locked away in the soil. As soluble P is used by plants and microbes, or is washed away by rain, more stable P is converted to labile P and finally to soluble P.

Insoluble forms of phosphorus are found in bone meal and rock phosphate, which are both sold as fertilizer. These forms of phosphorus break down very slowly. In neutral and alkaline soil, rock phosphate can take 100 years to break down, making it a poor choice for adding phosphorus.

Some agricultural soils have low levels of phosphorus because they have been heavily harvested for many years, but this is not true of most home gardens. There are two possible exceptions. First, very sandy soil has few nutrients of any kind and may benefit from the addition of more phosphorus on a regular basis, based on actual soil tests. The second exception is new homes built on land that was intensely farmed. Adding phosphorus for a couple of years will replenish the P level taken out by agriculture. Once this is done, little or no P needs to be added in future years.

Soil Myth: Phosphorus Stimulates Root Growth

Early studies on fertilizer showed that a lack of phosphorus resulted in poor root growth. This led to the idea that high levels of phosphorus would stimulate root growth, and suppliers started marketing products like transplant fertilizer and root boosters. Bone meal became very popular since it contains plenty of phosphorus.

Scientists soon discovered that roots grow just fine provided the soil contains an adequate amount of P. Excessive amounts actually slow down root growth.

Unfortunately, the myth had spread, and many gardeners today still promote high levels of phosphorus at transplant time, or to gain more blooms. Unless you have deficient levels of P, which is unlikely in most garden soil, adding more will be of no benefit to plants, but it can pollute lakes and rivers.

Although P is critical for plant growth, too much can be toxic to microorganisms, especially mycorrhizal fungi, which in turn can affect plant growth.

Even governments and fertilizer companies have come to realize that phosphorus is not required by most home gardeners. Some US states have banned phosphorus in lawn fertilizer, and some companies have reduced or illuminated the level of phosphorus in garden fertilizers.

Because phosphate sticks very tightly to clay, it moves very slowly through soil, about ¼" (6 mm) per year. For this reason, phosphorus fertilizer is routinely placed deeper in the soil right next to the seed. Placing it near the surface will build up the P stored in soil but does nothing for the current crop.

The bottom line for home gardeners is simple: Too much phosphorus is bad for the environment and may harm soil organisms. Unless your soil test indicates a shortage of phosphorus, you should

Soil Myth: You Can Identify a Phosphorus Deficiency by Looking at Leaves

Assume the plant is showing classic phosphorus deficiency symptoms such as darker green leaves and purple or red pigmentation. This could be a phosphorus deficiency, but this can also be caused by a variety of cultural conditions, including cold temperatures, high light intensity, damage by pests, and a lack of water.

It could also be a nitrogen deficiency that reduces the plant's ability to absorb phosphorus, which then shows up as a phosphorus deficiency. Adding more phosphorus to the soil will not solve this problem since the plant needs more nitrogen.

In most cases, you can't determine a nutrient deficiency by looking at plant leaves.

assume that your soil has plenty of it and don't add more in the form of commercial fertilizers.

Phosphate, a form of phosphorus, has historically been used in soaps and other cleaning products. Phosphate pollution from both fertilizer and soap is a main cause of contamination in lakes, where it causes algae blooms, which in turn reduce the amount of oxygen in the water, thereby killing fish. This is one of the main reasons that phosphate has been removed from soap products in recent years.

It is necessary for everyone to reduce the use of phosphorus fertilizer and apply it only when a soil test indicates it is needed.

Potassium

Potassium is not often used as a building block for large molecules, but it is critical for controlling the movement of CO_2 and water by regulating the opening and closing of stomata in leaves. It is involved in the formation of ATP, proteins, and starch and activation of enzymes responsible for plant growth. It also plays a key role in photosynthesis and preparing plants for a cold winter.

Because potassium is clearly a major nutrient, it is logical to think

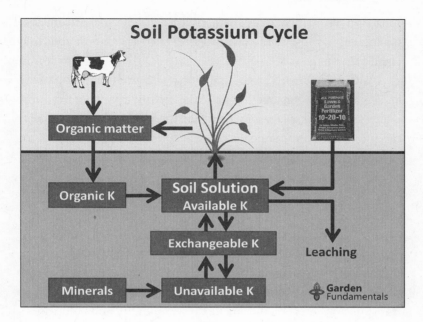

Soil potassium cycle.

that a plant needs a lot of it, but that is not true. Plants recycle potassium, using the same molecule over and over again to control things, with very little being depleted in the process.

Potassium salts dissolve easily in water and are absorbed as ions by plant roots. The potassium cations stick to clay and organic matter in a form called *exchangeable K*. As plant roots remove the portion in the soil solution, more potassium will move from clay and OM into the soil solution, becoming available to plants.

Potassium moves faster through soil than phosphorus but not nearly as fast as nitrogen. This means that the potassium in fertilizer sprinkled on top of the soil takes quite a while to reach plant roots, but once there, it is available for a longer time.

As much as 95% of the potassium in soil is locked up in minerals as *unavailable K*. Except for very sandy soil, most soil contains lots of potassium and is rarely deficient.

Potassium is not incorporated into large organic molecules. Instead it exists as free ions in the plant juices that occur in and around cells. As a result, most potassium is released fairly quickly from dead organic matter, and aged organic matter such as compost adds only small amounts to soil.

Plant roots absorb potassium along with water and don't have an active system for taking it in. If the soil solution contains more potassium, plants will absorb more, and therefore plants can possibly take up more than they need. There is no evidence that such an excess benefits plants—they don't get stronger or more cold resistant by overfeeding with potassium.

Calcium

Next to N, P, and K, calcium is the most used nutrient by plants. Its main purpose is to provide structural support to cell walls.

Roots absorb calcium passively at the same time they absorb water. Therefore, the amount absorbed by a plant depends on its transpiration rate. Conditions like high humidity and cold will reduce the amount of transpiration, thereby reducing this amount.

Except in highly acidic soil, calcium deficiencies are not common, but high levels of potassium can restrict the amount of calcium absorbed, leading to apparent deficiencies.

The calcium cations are absorbed by clay particles, where they help stabilize soil structure.

Magnesium

Although magnesium behaves very much like calcium in soil, it is more easily leached than calcium, and deficiencies are more common. Except for sandy soil and acidic soil, most contain adequate amounts.

Magnesium plays a critical role in photosynthesis and activates enzymes. Indirectly it is responsible for many reactions taking place in a plant.

Plants absorb magnesium as an ion in both a passive way and by active diffusion. The amount found in plants is directly related to the amount in the soil solution. Plants are able to mobilize magnesium, unlike calcium, and move it from older leaves to younger ones.

Potassium and ammonium compete with magnesium, and they can reduce the amount of magnesium uptake. Excess additions of either can result in an apparent magnesium deficiency.

Sulfur

Sulphur is not normally considered to be one of the main nutrients, but it probably should be. That is why some countries are starting to show the value of fertilizer as NPKS, where the S stands for the percentage of sulfur.

A plant uses sulfur and phosphorus in equal amounts. Sulfur is an essential component of proteins and various oils. The flavor and odor of onions is due to sulphur-containing oils. It also regulates important processes such as chlorophyll formation, and it plays a critical role in the creation of root nodules in legumes that are responsible for fixing nitrogen gas.

Like nitrogen, sulfur moves quickly through the top layer of soil, but then it gets trapped in the subsoil. Plants with longer roots are able to use them to reach this sulfur. Because soil tests are normally performed on the top layer, they are not very useful in determining the total amount of sulfur available to plants.

Unless it's sandy, your soil probably has enough sulfur.

Free sulfur originates mostly from the weathering of sulfate minerals. Not including minerals, 70% to 90% of the sulfur is tied up in organic matter as large molecules.

In recent history, sulfur deficiencies were quite rare since enough was supplied by acid rain and impure fertilizers. The move to less pollution and more highly refined fertilizers is resulting in an increase in reported deficiencies in agricultural soil. Home landscapes are rarely deficient.

Micronutrients

Plants use very small amounts of micronutrients, but that does not mean they are any less important. If a plant can't get an essential nutrient, it won't grow. Micronutrients are available in sufficient amounts in most soil, with the exceptions of very sandy soil and soil with extreme pH. Unless you know of a problem, assume these are not deficient in your soil.

Cation Exchange Capacity (CEC)

Because clay and organic matter are negatively charged, they hold onto positively charged ions (cations). Many of the nutrients critical to plant growth are cations, and therefore, the ability to hold them in soil is a key property.

Cation exchange capacity is a measure of the soil's ability to hold cations. A higher CEC means that the soil can hold more cations, which is good for soil fertility. A lower CEC means that there are fewer cations, and if the value is low enough, it will result in poor plant growth.

CEC also affects the stability of soil structure and soil pH since the hydrogen ion is a cation. High-CEC soils have a greater water-holding capacity, and acidification is more difficult.

The CEC varies depending on the percentage of clay content, the type of clay, soil pH, and the amount of organic matter. Pure sand has a very low CEC because the soil particles are not charged (CEC = 2). Different types of clay have CEC values ranging from 10 to 100. Organic matter has CEC values of 250 to 400. These numbers clearly illustrate the importance of clay and organic matter to soil fertility.

The common unit for measuring CEC is mEq/100 g, which is equivalent to centimoles of charge per kilogram. Both units are used on soil test reports.

High-CEC soils absorb not only nutrients but also other chemicals such as herbicides. If these are being sprayed on soil, you might need a higher level of spray since some will be lost as it sticks to soil particles.

Less fertilizer applied more frequently may work better in low-CEC soils since the soil does not hold onto the nutrients and leaching occurs more rapidly.

CHAPTER 3

Soil Life

There is more biodiversity in the top foot of soil than anywhere else on Earth. It is a difficult environment to study, and to date, we know very little about the organisms that live there, and we know even less about their interactions.

To the general public, soil is some mysterious black stuff that helps plants grow. They see the occasional ant or earthworm due to their size, but the majority of life is too small to be seen. Gardeners are aware that microbes are essential, but many can't clearly articulate why. I hope to change that in this chapter, which will have a close look at the significant organisms and explain how they contribute to soil fertility and plant health.

Energy Food Web

All living organisms require energy to live. Organisms that use the sun as an energy source are called *autotrophs*, and those that use carbon compounds are called *heterotrophs*. A few bacteria get their energy from minerals and are also called autotrophs.

Plants, algae, and a few specialized bacteria are autotrophs, and they can produce their own carbon food through photosynthesis. They use the energy of the sun to convert CO_2 from the air into energy-rich compounds containing carbon.

The heterotrophs can't make their own food and rely on eating organisms that contain carbon-based compounds such as sugars,

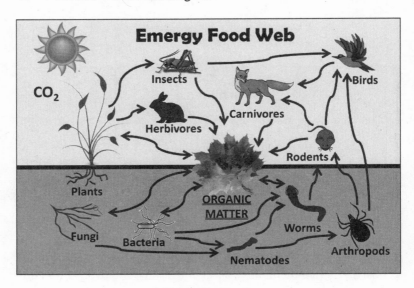

Energy food web.

fats, and carbohydrates to provide the energy to carry out all of the chemical reactions needed to survive. Eaten food is digested to release these chemicals that are then distributed to most of the cells in the body. Carbon is our energy source.

The energy food web diagram shows how carbon moves from plants to most other types of life forms, passing from one to another, until the organism dies. At that point, the carbon in the dead organic matter is used by microbes as an energy source, and they become the food source for larger heterotrophs. The carbon then moves from organism to organism until it again becomes dead organic matter, and the cycle repeats. The key point is that all of the carbon energy originates with the autotrophs, and plants are major contributors.

If the energy food web continued as I have described it, the amount of carbon compounds on Earth would be continuously increasing, but that is not what happens. As organisms digest and use the food, they produce CO_2 through respiration. Animals get rid of this excess CO_2 by breathing it out; plant roots respire and expel CO_2. A significant amount of CO_2 is also created during composting, as a result of microbe respiration.

The Power of Large Molecules

If we have a close look at a single cell, we find that it is full of all kinds of chemicals. Humans are about 60% water, and plants can be as high as 95% water, with most located inside cells. Cells also contain nutrient ions that are floating around in the water ready to be used by the cell to form new larger molecules.

A significant part of a cell consists of molecules such as proteins, carbohydrates, fats, and DNA. To put the size of these molecules into perspective, a protein known as ATP synthase is 10,000 times bigger than a CO_2 molecule. DNA is thousands of times bigger than this protein.

The contents of the cell are enclosed in a membrane, a kind of outer shell that keeps the insides in place. Inside, molecules are constantly being broken down into small molecules, only to be built up again into new large molecules. At any given time, most of the nutrients are in the form of large molecules.

The process of converting nutrients into larger molecules is called *immobilization*. The reverse process is called *mineralization*, whereby large molecules are broken down into simple minerals and ions. When an organism grows, it makes new cells by taking the basic building blocks, the nutrients, and immobilizing them into a vast number of large molecules.

What happens when an organism dies? The cells stop working, and the molecular activity mostly stops. Molecules are no longer created or destroyed. Whatever was there at the time of death tends to remain unchanged, but some small changes do take place. A banana that has been sitting on the counter for a week still looks like a banana, it still has most of the cells intact, but it's starting to go brown. To our eyes, these changes seem to be significant, but on a molecular level, they are very minor. Almost all of the large molecules remain unchanged even after you throw the banana into the compost bin.

Plants can absorb only nutrient ions and a few small molecules through their roots. If you place a rotten banana next to some plant roots, it will do almost nothing for the plant. The plant can't use the large molecules, and the nutrient ions are still being held inside the protective cell walls.

Microbes, on the other hand, are able to use the banana as a food source. They release enzymes that attack the cell walls. Once the cell wall is breached, the insides run out, much like with a broken egg. The released water and nutrient ions can now be used by plants, but the majority of material is still in the form of large molecules and is useless to plants.

Slowly, many different types of microbes enter the picture and start decomposing the large molecules into small ones and then into even smaller ones. At some point, all of the large molecules are mineralized into nutrient ions and other small molecules like carbon dioxide and water. In both a compost pile and in soil, this happens very slowly over several years.

This story reveals one of the special characteristics of organic matter. It decomposes very slowly, all the while releasing plant-useable nutrients. The large molecules found in both dead and living organic matter act like a storage container for plant nutrients. This storage mechanism is critical to the way nutrients are cycled and made available to plants.

Ratio of Fungi to Bacteria

Bacteria and fungi are critical for the movement of nutrients in soil and consequently play a major role in plant health.

The ratio of fungi to bacteria varies depending on the location and history of the soil. Deciduous forests can have a biomass ratio of 10:1, while coniferous forests have a ratio of 100:1. These are considered fungal environments because the mass of fungi is so much larger than the mass of bacteria. Natural grasslands and highly productive agricultural lands have rations of 1:1 or less, indicating that bacteria are more dominant. Highly cultivated agricultural land will have a much lower ratio since cultivation destroys fungal hyphae.

This difference has led some to speculate that managing and modifying the ratio is good for growing specific types of plants. There is little scientific evidence that a specific ratio is required to grow a healthy plant, and there is no support for the idea that the ratio can be altered by adding either fungi or bacteria to the soil. Building healthy soil will increase the number of both fungi and bacteria, which is important. The actual ratio is not significant.

Chemicals in the Soil

All kinds of chemicals are naturally added to soil. Plants, microbes, and animals all excrete thousands of chemicals, and even more are added when they die and decompose. Plants produce a large number of natural pesticides that are added to both air and soil. You might think that since they are natural, they are all safe, but that is far from the truth. Many of these are toxic or carcinogenic to a variety of life forms if they accumulate in high enough concentrations. Human-made chemicals are also added to the soil in the form of air pollution falling to Earth, the spray of pesticides, and the addition of drugs from such things as manure.

All of these chemicals would slowly build up and make our soil toxic for growing anything were it not for the microbes. They are able to absorb or decompose just about every chemical added to soil. The exceptions include heavy metals like mercury and cadmium, which are elements that are already decomposed as far as possible. Microbes essentially eat up all of these chemicals. There are even bacteria that consume Roundup and oil. Over time any chemicals that are added to soil are broken down into basic nutrients, CO_2, O_2, and water.

Scientists talk about the half-life of a chemical, the time it takes for one-half of the chemical to be naturally removed from the environment. Glyphosate, the active ingredient in Roundup, has a half-life of about 3 months in soil. DDT, however, has a half-life of 2 to 15 years, depending on conditions. Some pesticides are degraded so quickly by bacteria that an excess needs to be applied to make up for the amount lost to bacteria.

Pathogen Control

The majority of microbes in soil are good guys and go about their lives with little interference with plant growth. A very small number are plant pathogenic microbes that cause diseases. These pathogens exist everywhere and are constantly present in soil. It is amazing that all plants are not sick.

The reason plants are not infected more is that there is tremendous competition in soil. It is a dog-eat-dog world down there. Microbes are continuously waging war against each other, not only for space but also for resources, like food. They also carry out chemical warfare by excreting a wide range of toxic chemicals.

A healthy, diverse population keeps any one microbe from taking over the whole community, and since the non-pathogenic microbes outnumber the pathogens, the pathogens are kept in check so they can't get a foothold on plants.

As we'll discuss later, in the chapter on the rhizosphere, plants cultivate the good guys and encourage them to live next to their roots. This layer of microbes acts like a shield, making it very hard for pathogens to get close enough to cause problems.

Identification of Microbes

Studying soil life is extremely complex, and we are just starting to scratch the surface. Many microbes are very similar, making identification difficult. Many can't be cultivated in the lab, adding another layer of complexity. The sheer number of species is staggering.

Consider these facts:

- A teaspoon of soil contains more microbes than people on Earth.
- It is estimated there are 1 billion soil bacteria species, and only 30,000 have been identified.
- Earth could contain a trillion microbe species, many of them living in soil. A very small fraction of these have been identified.
- There are 1 to 5 million species of fungi in soil; only a small fraction has been identified.

All of these numbers are estimates, and scientists are not in agreement on the numbers. We know so little about soil life that this is all a big guess.

CHAPTER 4

Bacteria

Bacteria are probably the most important life form in soil, and without them, plants could not grow. They fix nitrogen from the air, decompose dead organic matter, extract minerals from rocks, and help soil aggregation. A good understanding of them provides insight into why many gardening techniques work or don't work.

Bacteria are very small; 500,000 take up no more room than the period at the end of this sentence. A single teaspoon of soil contains as many as 30,000 species. The sheer number of bacteria in soil is astounding. A gram of fertile soil, about the weight of a paperclip, contains up to a million bacteria. It is hard to get your head around such a big number, but consider this: there are as many bacteria in two teaspoons of fertile soil as people on Earth.

They are single-celled and have a variety of shapes including spheres, rods, and spirals. When it is time to reproduce, they split into two cells, each being a copy of the original. In perfect conditions, this can happen every 20 minutes, producing huge amounts of bacteria in a short time. In a lab situation, a single bacterium can multiply to 5 billion in 12 hours.

We tend to think of bacteria as being one type of organism, but the diversity is huge. There are species that live in virtually every type of environment on Earth and eat almost everything, even spilt oil and jet fuel. Some like it hot, some like it cold, some wet, some dry. They are found everywhere, but each environment hosts a different community of species.

Most bacteria are heterotrophs, which are organisms that get their energy source from carbon. A few are autotrophs, which get their energy from soil minerals. Aerobic bacteria use oxygen in the same way we do. Another group, anaerobic bacteria, prefers environments that have a very low level of oxygen, and there is even a group that can live in both environments. An example of this latter group is E. coli, a bacterium that lives in our intestine, which is anaerobic, as well as in soil, which is usually aerobic.

Heavy clay soil containing very little organic matter tends to be anaerobic because oxygen can't get into the soil. When you dig in this type of soil, you can smell the foul odor that is produced by anaerobic bacteria. The smell of rotten meat, rotten egg gas, vomit, and vinegar are all smells produced in anaerobic conditions. Anaerobic bacteria can also produce alcohol, which is toxic to plants. On the other hand, the smell of fresh soil in a forest is produced by aerobic bacteria.

Bacteria can move in various ways. Some have flagella, a kind of tail that acts like a propeller to move them through the environment. For their size, they can move quite far, as much as 5 micrometers in a lifetime. A human hair has a thickness of 20 to 200 micrometers.

What Do They Eat?

Bacteria don't have mouth parts, so they don't really eat, but they do need to get nutrients. They have a cell wall that keeps important chemicals in and keeps unwanted chemicals out. Bacteria use something called *active transport* to selectively move molecules through the cell wall. Think of it like the border around your country. There are controls in place that keep unwanted things out but let others in. A bacterium wall works the same way. It lets nutrients in and keeps harmful molecules out.

This still leaves bacteria with a problem. Assume they are sitting next to some fresh plant material. All of the nutrients in this material are tied up in large molecules that are too large to transport across the cell wall. Instead of just waiting for something to happen, bacteria take aggressive action. They excrete a variety of enzymes through their cell wall. These enzymes start to decompose the organic matter, breaking the large molecules into smaller and smaller molecules until the nutrient ions are released. The bacterium then absorbs these nutrients. Once inside the bacterium, the nutrients are immobilized into new large molecules.

I have described the process in simple terms, but it is actually quite complex. The bacterium does not have eyes or a brain, so it doesn't know if there is organic matter nearby. The process is further complicated by the fact that enzymes are fairly specialized, and each one decomposes only a particular type of large molecule.

To solve these problems, bacteria excrete a wide range of enzymes, hoping that some organic matter exists nearby and that it contains the right kind of large molecules for the excreted enzyme. Once a nutrient molecule is released, it tends to just sit there, or flow along with the water. The poor bacterium has to wait until the nutrient molecule hits its cell wall, and then it can grab it and pull it into the cell. The process seems very inefficient, but it must work reasonably well because bacteria thrive in soil.

This activity—making enzymes, transporting molecules through cell walls, and dividing to reproduce—all requires a lot of energy. That energy comes from carbon-containing molecules, such as sugar, that the bacterium must also absorb from its environment. This food source is as important as nutrients.

Organic matter has an abundance of molecules containing carbon, and these are released in the decomposition process. Almost 50% of a plant is made up of cellulose, and bacteria enzymes can break this down into sugars that provide a good energy source.

Lignin is a major component in wood, and it too contains lots of carbon. Although bacteria can't decompose it, fungi can.

Where Do They Live?

Most bacteria live in the top 4 to 6 inches (10 to 15 cm) of soil, where there is lots of air, water. and food. The numbers fall off dramatically as you move deeper.

Bacteria live mostly in water because it is necessary to bring food molecules to their cell walls. As the water dries out of soil, they go into a form of hibernation called *biostasis* and wait until water is available again. They deploy a defensive mechanism to prevent drying out. Bacteria are coated with a biofilm made of sugars, proteins, enzymes, and DNA. This layer dries out much slower than pure water and protects them from desiccation.

Within soil there are various types of habitat. The large pores in soil don't hold too many bacteria because the water here quickly drains away, flushing bacteria with it. The very small micropores inside soil aggregates are a much better hiding place for them. These dry out very slowly, and the small size protects them from larger predators.

Large amounts of bacteria are found in the rhizosphere, the area right around plant roots that provides food, resulting in a high rate of reproduction.

Role in Disease Prevention

Bacteria can be either beneficial or detrimental to plants; luckily most of them are beneficial. A few do cause serious diseases, and symptoms include galls, overgrowths, wilts, leaf spots, soft rots, and scabs.

Some plants do have immunity to bacterial pathogens, but much of the prevention stems from bacterial competition. The enzymes that bacteria excrete digest both dead and living organic matter, including neighboring bacteria. Bacteria also produce toxic substances to kill other bacteria. Since the majority of bacteria are good guys, they tend to win these wars due to sheer numbers. Pathogens have a hard time developing a foothold in plants.

Larger soil organisms use bacteria as a food source and also play a major role in controlling pathogens. The key to healthy plants and pathogen control is to maintain a large number of microbes that show a wide diversity.

Ideal Environment

Since bacteria are good for both plant health and creating highly aggregated soil, it is your job to keep them happy. They like the basics in life: oxygen, food, water, and the right temperature. The oxygen and water part is fairly simple and is no different from the requirement for plants. Soil needs to have around 25% air and adequate levels of moisture. Pathogenic bacteria tend to prefer anaerobic conditions, making the presence of air even more critical.

Almost any organic matter can serve as a food source. Dead plant roots, plant refuse, and other dead organisms play an important role here, but you can help out by adding more organic matter. Any gardening practice that increases the amount of organic matter or reduces the loss of organic matter will benefit all microbes in the soil.

The last requirement is temperature, and it's mostly out of your control since it is a function of location and environment. The good news is that microbes are fairly flexible with this. They all have a preferred range of temperature, but they are also able to go into biostasis if conditions are not ideal.

The types of bacteria that are active in spring and fall will be quite different from those active in summer. These changes in population happen all of the time, and gardeners don't really have to worry about them.

Role in Building Soil Aggregates

Bacteria are coated with a biofilm comprising a mixture of sugars, proteins, DNA, and all kinds of other chemicals. This protects them, but it also produces a side benefit for soil: it helps stick small soil particles together and plays a major role in aggregation.

Each bacterium is very small and has little effect on soil, but the vast total quantity has a substantial effect on the soil structure. Also significant is the fact that bacteria are relatively short-lived. When they die, this biofilm is left to coat soil particles.

Conditions that Harm Bacteria

The life of bacteria is a hard one. They are quite small and can't move very far. If the local food source runs out, they die. If the temperature changes too much, they go into hibernation. Almost every animal-like organism is larger and wants to eat them. Even other bacteria excrete enzymes to try and digest them. If it were not for their vast numbers, they would probably not survive.

Food is critical for them, so anything that reduces the amount of organic matter will also reduce their numbers. Tilling increases the amount of oxygen in the soil, which speeds up the decomposition process and in turn reduces the amount of organic material available. Short term, this increases bacterial populations, but in the long term, it decreases them. Soil activity that causes compaction or reduces oxygen levels will decrease the number of bacteria.

Soil Myth: Fertilizer Kills Bacteria

There is lots of misinformation about what synthetic fertilizer does to bacteria and other soil life. The idea that it kills microbes is completely false. There is absolutely no difference between nutrient ions coming out of a fertilizer bag and those released by an organic source. Because they are identical, and bacteria need them to sustain life, it is clear that synthetic fertilizer does not harm bacteria. Bacteria populations normally explode after the application of fertilizer because of the sudden increase in food. Too much of a good thing can be harmful, but that applies to synthetic fertilizer as well as organic matter.

Plant diversity is also an important factor that affects bacteria. A higher diversity of plants results in a higher diversity of bacteria. The monocultures found in vegetable gardens and agriculture reduce bacteria diversity.

Nitrogen Fixation

Nitrogen is added to soil from the decomposition of special minerals, lightning, and the fixation of nitrogen. Lightning converts atmospheric nitrogen into nitrogen oxides (NO and NO_2), which can then be converted into the ammonium and nitrate ions. This is an important process, but it is not the most significant source of soil nitrogen.

Most of the useable nitrogen in soil is produced by bacteria. Specialized nitrogen-fixing bacteria are able to take the nitrogen gas and turn it into ammonium and nitrate ions. These bacteria can be free-living or be attached to plant roots in a symbiotic relationship.

Chemical manufacturing can also fix nitrogen, producing synthetic fertilizers containing ammonium, nitrate, and urea. This is now the major source of plant nitrogen in developed countries.

Free-living Nitrogen-fixing Bacteria

These bacteria live in soil and are not directly associated with particular plants. Most of them need a carbon source for energy, usually dead organic matter. A few types can use minerals as an energy source.

These bacteria produce an enzyme called nitrogenase that fixes nitrogen outside their cell walls. Because oxygen inhibits this enzyme, most of these bacteria live in anaerobic conditions. In natural situations, these organisms are not a major contributor to the total amount of fixed nitrogen, but they can be important is special environments. For example, some species tend to live in the rhizosphere of grasses and cereal crops, where they contribute a significant amount of nitrogen.

Symbiotic Nitrogen-fixing Bacteria

A specialized group of bacteria form symbiotic relationships with specific plants. The best known of these is the association of *Rhizobium* bacteria with legumes, such as clover, alfalfa, soybeans, broad beans, and peas.

The plant initiates contact by releasing flavonoids through their roots. The bacteria sense these compounds and attach themselves to specialized root hairs. The legume then forms a growth around the bacteria, a nodule that not only protects the bacteria but also provides an anaerobic environment that is critical for nitrogen fixation.

The plant feeds the bacteria by excreting sugars and other nutrients into the nodules. The happy bacteria then fix nitrogen to form ammonia, which is converted to nitrate as it is absorbed into the plant. You can think of these nodules as being little factories that make nitrogen for the plant.

This ability of legumes to have their own nitrogen source means that they are able to grow in environments that have very low natural nitrogen levels, making them very competitive. But there is a cost to the plant for this nitrogen. The plant may use as much as 20% of its photosynthate (sugars produced by photosynthesis) to maintain the bacteria.

There are numerous species of *Rhizobium*, and they are specialized for various legumes. For example, peas and beans are infected by different species. In order for legumes to form nodules and host the bacteria, the bacteria must be present in the soil. If your soil does not contain the right strain, no nodules will be formed.

Gardeners solve this problem by inoculating seed with the right bacteria at the time of planting. Little packs of bacteria can be purchased from seed companies, and you can buy seed that is already coated with the right bacteria. Once the bacteria is in the soil, it will survive there for several years, so even a four-year crop rotation does not need to be inoculated each time.

How do you know if you have the right bacteria in the soil? Grow the legume and have a look at the roots halfway through the summer or in early fall. You can easily see the pea-sized nodules if they are there. They are most visible as plants bloom.

If the plant did not make nodules, you either do not have the right bacteria in the soil or have too much nitrogen. Excess fertilizer will prevent the formation of the nodules, since the plant simply does not need the bacteria.

Frankia Actinomycetes

Actinomycetes, or filamentous bacteria, are important microbes that have characteristics of both fungi and bacteria. Frankia form symbiotic associations with non-leguminous plants, mostly trees and shrubs such as alders, sea buckthorn, and *Casuarina* species. Many of these plants are pioneer species that grow in very poor soil.

CHAPTER 5

Fungi

The fungi in soil are as important as bacteria, and they also play a critical role in plant health. They are decomposers that are critical in the decomposition process of organic matter carrying out functions that bacteria are unable to do. Specialized mycorrhizal fungi extend the reach of plant roots as they form a symbiotic relation with them, allowing them to reach into very small pore spaces.

Fungi are a strange group of organisms. They have some plant-like characteristics, such as the hyphae that function like roots and grow out into soil looking for nutrients. But fungi have no chlorophyll, so they can't make their own food. They depend on carbon compounds for their energy source just like animals. When conditions are right, they flower, or at least make a mushroom growth, referred to as a fruiting body, that looks and functions similar to a flower.

Beer, wine, and bread are the result of the single-celled yeast fungi, and some products including probiotics include fungi. The mold on old bread and cheese is also evidence of fungi. When you eat mushrooms, you are eating the fruiting bodies of multi-celled fungi.

Fungi start life as a spore, a special kind of seed that is normally produced by the fruiting body. The fine powder given off by an old puffball consists of spores. If you take a mushroom cap and let it sit for a day, you will find the powdery spores under it. Spores are very small and easily float through air. The breath you just took sucked in a bunch of them. They travel great distances on wind currents, which means that fungal spores are everywhere. They exist universally in all climates and can even be found in the Antarctic.

When these spores land in a suitable environment, they sprout and start growing hair-like hyphae (hypha is singular) made up of long chains of cells that form a type of tube. Liquid, nutrients, and other compounds flow easily from one end of the tube to the other. The tip of the hyphae is continuously growing and branching to form new arms of hyphae that are all connected. They can grow as fast as 40 micrometers per minute, about the thickness of a human hair. These hyphae tend to mass together to form larger clumps, which are called mycelium. When hyphae die, they leave a vast network of tunnels that allows water and air to move through, as well as other smaller organisms.

Hyphae are 1 to 3 micrometers in diameter, which are too small to be seen, but if you find white, fuzzy, cottony material in soil, you are looking at a clump of mycelium. The white coating on leaves caused by powdery mildew is due to fungal hyphae.

A single gram of soil, the weight of a paperclip, can contain 100 to 1,000 meters of hyphae. Fungi are less numerous than bacteria, but since each organism is larger, the total mass of fungi in soil is about the same as its mass of bacteria, and together they account for much of the microbe population.

Over 100,000 species have been identified, of which 70,000 can be found in soil. Their estimated total is 1.5 million. We still have much to learn about soil fungi.

Fungi are able to condition the environment around them. They exude acids to make it more acidic, and they produce a wide range of chemicals, such as antibiotics that help control parasitic organisms that use fungi as a food source. Penicillin is probably the best known fungal produced antibiotic.

What Do They Eat?

Fungi are heterotrophs, which mean they cannot make their own food and need to find a carbon source, which they do much like bacteria. They excrete various enzymes and other compounds that decompose the large organic molecules surrounding them. Once converted to small nutrients and sugar molecules, they can be absorbed and transferred several feet along the hyphae to where they are needed.

Unlike bacteria, some fungi can digest the really tough organic matter like lignin and cellulose. The tip of hyphae can produce enzymes that let it penetrate hard surfaces like plant leaves and bits of stems to get at the more digestible material inside. Most of the absorption of food takes place near this tip.

A unique quality of fungi is their ability to grow hyphae above the top of the soil to penetrate leaves and other plant refuse laying on the surface of the soil. They are critical for cleaning up the plant litter that drops in fall. The nutrients in this material are then moved deeper into the soil for other microbes and plants to use.

This above-ground growth becomes very evident if you pile up leaves. They don't contain enough nitrogen for bacteria to decompose, but fungi soon take on that role and produce leaf mold. Leaf mold is a combination of partially decomposed leaves and lots of mycelium, making a great addition to any garden.

Wood chip mulch is also invaded by fungi. After several weeks, these pieces are stitched together with fungi hyphae that are slowly decomposing the wood. Many varieties will even form harmless fruiting bodies on it. Google crazy-looking fungi like the dog vomit fungus, stinkhorn mushroom, and dead man's fingers to better understand this bizarre world.

Where Do They Live?

Fungi are aerobic and do not need a water film to survive. They are able to grow in the air gaps in soil and travel quite some distance to find food.

Hyphae are thinner than plant roots, but thicker than bacteria. They travel through pores in the soil that are bigger than the hiding places of bacteria, but they are able to get into openings that are too small for plant roots.

Fungi at War

Some fungi are able to attack living prey. About 150 different species are able to capture nematodes, either with sticky adhesive pads or a noose-like loop that snares and captures them. After capturing them, the fungi exude enzymes to digest their living prey.

The oyster mushroom is a common delicacy. but it is also a vicious hunter. They grow on trees and dead fallen wood. The wood provides them with a high-carbon source, but it lacks the nitrogen needed to fully digest it. Nematodes make a good nitrogen source.

To set a trap, the oyster mushroom fungus exudes chemicals that smell like diner to a nematode. When the nematode gets close enough, it bumps into specialized filaments that are tipped with a toxin that paralyzes them. The fungus then grows hyphae into the nematode and digests it from the inside out.

Fungi are not without their own predators. They are attacked by a variety of organisms, and their main line of defence is chemical warfare. They exude toxins, insecticides, and antibiotics that keep predators away, prevent them from developing correctly, and even kill them. After all, dead predators make a good meal.

The fruiting bodies can also be toxic; if you eat the wrong mushroom, the fungi can kill you.

Fungal Parasites

The majority of fungi in the soil are good guys and beneficial to plants. Unfortunately, there are also some bad guys that parasitize plants. You might know them as mildew, root rot, rust, damping off disease, botrytis, scabs, cankers, and fusarium. Fungi cause more harm to plants than other soil organisms.

In many cases, they don't kill the host plant, but they do weaken it. In agricultural crops, it reduces the yield; in horticultural plants, it can reduce vigor and flowering. Soil-borne fungi tend to invade roots and form some kind of deformed growth, such as club root on cabbage.

Fighting these diseases needs to be done on an individual basis. Identify the problem, and then research specific ways in which to reduce the issue. There is no general cure that works for all fungi problems.

Mycorrhizal Fungi

Mycorrhizal fungi are a special group of fungi that form a very close symbiotic association with plants. It is estimated that 95% of all plants form such a relationship.

These fungi have been divided into two categories: ectomycorrhizae (outside roots) and endomycorrhizae (inside roots). Ectomycorrhizae are less common, but they are important for woody plants. An individual tree may have associations with 15 or more different fungi at the same time. There may be as many as 20,000 species in soil. The hyphae of these fungi wrap around roots, forming a type of sheath that does not normally penetrate the actual root. Once the sheath is formed, they are able to exchange chemicals with the roots. This group forms many of the noteworthy mushrooms, including amanitas, chanterelles, and the prized truffles.

Endomycorrhizae are more widely spread and form associations with many plant families. Their hyphae actually penetrate and grow inside the roots. This fungi group is further divided into arbuscular mycorrhizae (AM) and ericoid mycorrhizae (EM), depending on how they grow. The AM fungi are major partners for most agricultural crops, and scientists have identified 240 species, but genetic diversity is much greater than this.

All mycorrhizal fungi form a symbiotic relationship with plants that benefits both. The fungi receive sugars and other nitrogen sources from the plant in exchange for water and mineral nutrients. The plant shunts about 15% of its food to the fungi.

An interesting association is formed with the Indian pipe (*Monotropa uniflora*), also known as the ghost plant or corpse plant. This is a white mushroom-like herbaceous plant that has no chlorophyll. It forms an association with mycorrhizal fungi that in turn forms an association with trees. The fungus shunts sugars from the tree to the Indian pipe plant.

The fungi hyphae can extend the effective root system of a plant by a factor of 1,000, increasing its ability to access nutrients. The hyphae are also much thinner than roots, so they are able to collect nutrients from small pores that exclude roots. Fungal enzymes and acids also free nutrients like copper, calcium, magnesium, and zinc from rock material.

Phosphorus levels are always low in the soil solution, but fungi are able to pick up significant amounts due to their large surface area. This is one of the most important nutrients they provide to plants. It should also be noted that large amounts of phosphorus in soil, from too much fertilizer, will inhibit mycorrhizal growth.

These fungi also protect plant roots from disease and can absorb toxins before they reach the roots. Natural soil contains lots of different species, but cultivation, removal of topsoil, and compaction reduce diversity, making it more difficult for the fungi to grow.

CHAPTER 6

Other Organisms

There are many kinds of organisms in soil. The following are some of the more important ones for soil health.

Actinomycetes

Actinomycetes, called mold bacteria and thread bacteria, look and grow more like fungi but biologically are similar to bacteria. They grow hyphae-like threads that consume resistant organic matter, and they are tolerant of dry soil, alkaline soil, and high temperature conditions. They produce chemicals that stop the growth of other microbes, such as streptomycin and actinomycin, which are now commercially available drugs.

Actinomycetes tend to be found in decaying organic matter. They can protect plant roots from disease, and in a few cases, they cause diseases such as potato scab. Frankia, a type of actinomycete, forms symbiotic nitrogen-fixing associations with over 200 species of plants and are responsible for the earthy smell of damp, well-aerated soil. Their affinity for higher temperatures and ability to decompose tough organic matter makes them an important element in hot composting.

Algae

Soil algae are single-celled organisms that photosynthesize like plants and can be green, blue-green, or brown. A gram of soil can

Microbe Populations in Soil.

Organism	Number in 1 gram of healthy soil
Bacteria	1,000,000,000
Actinomycetes	100,000,000
Fungi	1,000,000
Algae	100,000
Protozoa	10,000
Nematodes	1,000

contain 10,000 to 100,000 algal organisms. Some live on their own in soil, and others live in combination with fungi to form lichens, which produce acids that slowly dissolve the rock on which they live.

Algae must be exposed to some light and tend to live in the upper level of soil. They also need a fair amount of moisture to grow well. The blue-green algae, also called cyanobacteria, are able to fix nitrogen, but their contribution to soil nitrogen is small.

Protozoa

Protozoa are not technically animals, but it is useful to think of them as such. They are single-celled and move around the soil solution looking for prey, which consists mostly of bacteria. You might be familiar with the protozoa called amoeba, which is commonly studied in school.

Some have a mouth-like aperture for sucking up prey, but most simply engulf their prey. Once inside, their digestive enzymes convert food into smaller molecules. They can eat up to 10,000 bacteria a day, and the larger protozoa eat the smaller ones. They can even eat and control pathogenic nematodes. Some protozoa contain chloroplasts that can photosynthesis to produce sugars.

The vampire amoeba, vampyrellids, chews perfectly round holes in fungi and uses them as a food source. They do this by attaching to the surface of the fungi and excreting enzymes that digest the hole. The amoeba then sucks up the fungal cells.

Protozoa live in a film of water in the top 8 inches (20 cm) of soil. If it gets too dry, they form a cyst and hibernate until moisture returns. Some species can survive up to a year in this condition.

Their size ranges from 5 to 1,000 micrometers, compared with 1 to 4 for bacteria. They are too big to inhabit the micropores like bacteria, so they hang out around soil pores waiting for bacteria to emerge.

Protozoa are a critical part of soil since they keep bacteria numbers in check. Bacteria also contain a lot of nitrogen with a C:N ratio of about 5:1. Protozoa need much less nitrogen and have a ratio of 10:1. The excess nitrogen is excreted into the soil as ammonium, which plants and other organisms can use. This nitrogen is a major source for plants.

There are 1,000 to 1 million in a teaspoon of soil, and about 250 of the estimated 30,000 species have been identified.

Some eat roots and damage plants, but most are not a pest in the garden.

Nematodes

Nematodes, eel-like worms that live in soil and around plant roots, can swim short distances in water films where they graze on bacteria, fungi, protozoa, and smaller nematodes. Some larger ones even attack insect larvae and are used to control grubs in lawns. A commercial nematode product is available for controlling snails and slugs (not available in North America).

Most nematodes are found in the top 6 inches (15 cm) of soil, where they can consume 5,000 bacteria a minute. Like protozoa, they need less nitrogen than the bacteria they eat and consequently excrete the excess into the soil solution for plants to use.

They are up to 2 mm long and 50 μm wide, and a teaspoon of soil contains about 50 to 500 of them. There may be as many as 40,000 species worldwide.

Nematodes can be beneficial or parasitic to plants. When they attack plants, they puncture roots and suck juices out, weakening the plant. These puncture wounds then provide an entrance for other pathogens. Roots respond by making root knots or swellings. A good example of this can be found in the root-knot nematode that affects carrots.

Arthropods

Arthropods include all insects and a wide range of insect-like animals including mites, millipedes, centipedes, pill bugs, ants, springtails, and termites. There are several million species on Earth. To provide some sense of scale, there are 1,000 ant species in North America, and that is just a fraction of the arthropods.

Many arthropods live in or on the soil at some point in their life cycle. Many feed on decaying organic matter, smaller animals, and fungi. Some, like the June beetle and Japanese beetle larvae, feed on plant roots. They can also be a major pest to the above-ground parts of plants.

They can have a significant effect on the soil by their digging and tunneling activities. This helps loosen soil and add large pores that are used by plant roots. They are also important shredders, chewing organic matter like dead leaves into smaller pieces that are then more accessible to microbes. Mites and springtails recycle 30% of the fallen leaves and woody debris in temperate zones.

Arthropods move around much more than the other life forms I have discussed, and they become the transportation vehicle for smaller organisms that catch a ride on them. In this way, they play an important role in maintaining a high degree of diversity.

Their relatively large size makes them visible to us, and we might think that they make up a significant amount of the living soil biomass, but nematodes and protozoa have a higher biomass.

Earthworms

Gardeners love to see earthworms because they represent a healthy soil. Environmentalists hate to see them since they are an invasive species in North America that are damaging wooded areas. We now even have the jumping earthworms invading soil and causing even more concern. Without doubt, earthworms have a major impact on soil.

Worms eat soil and organic matter that travels down a long digestive tube consisting of several key sections. The esophagus adds calcium carbonate as a way for the worm to rid itself of excess calcium. The food then moves on through the crop and into the gizzard. The gizzard uses swallowed stones to mash food into small particles. Enzymes are added to help digestion. The material continues into the intestine where fluids are added to further digest the food, and, like our own intestine, it absorbs nutrients.

This all sounds quite normal for an animal digestion system, but there is one other key ingredient: microbes. The worm controls moisture and pH levels to favor the growth of microbial populations, which play a major role in the digestive process.

Along with the soil and organic matter, worms also ingest large amounts of microbes. In fact, the microbes are their primary source of food, not the organic matter. They also excrete live microbes and play a significant role in distributing them throughout the soil.

The whole digestive system is not very efficient: only 5% to 10% of the ingested food is absorbed by the worm. The rest is excreted as vermicasts, or worm castings, mucus-coated particles containing undigested plant material, nutrients, soil, and a large amount of microbes. The microbial activity in worm casts is 10 to 20 times higher than it is in soil.

The worm is not really doing much composting. It does some digestion, but its main contribution to the process is that it breaks

Soil Test: Earthworm Count

As soil dries, earthworms move deeper into the soil, and counts become inaccurate. Counts are also lower early and late in the season. It is a good idea to test several times during the season and take an average.

Measure out 1 square foot of soil. Dig down 12" and remove all of the soil. Spread it out on cardboard or newspaper. Break up the soil, and count all of the worms you find. If your soil is healthy, you'll find at least 10 earthworms per cubic foot.

organic matter into small pieces and mixes it with microbes. A point that is not emphasized enough is that much of the composting process takes place after the casts exit the worm. This external processing may in fact be the most significant part since it is only then that nutrients are mineralized and made available to plants.

As worms burrow through soil, they create large pores, which are important for both aeration and for roots growth. They also move organic matter from the surface of the soil to lower levels. They prefer moist soil, a neutral pH, and mid-range temperatures. To the gardener, their presence indicates a healthy soil with plenty of organic matter and microbes.

They are a problem in forested areas because they speed up the decomposition of organic matter, and over time, they cause organic levels in wooded areas to drop. This will become a major environmental problem for trees and their supporting fungi.

There are about 7,000 species of earthworms, with 10 to 50 per square foot (1 to 5 per square meter) of soil. An acre can have 200 to 1,000 pounds of earthworms (230 to 1,100 kg/ha).

CHAPTER 7

Organic Matter

Every gardener knows that organic matter is important for soil, but few really understand why. It is much more than just a source of nutrients for plants. This chapter will have a close look at both the chemical and biological effects on soil.

All living things are based on carbon, and the term *organic* refers to this fact. Soil contains three basic forms of organic carbon, what some like to call the *living*, the *dead*, and the very *dead*. The living component, 15% of the total, contains all of the living organisms, including plant roots. The dead refers to any organic matter that is in the process of decomposition. The very dead material has completely decomposed and is in a more stable form, which includes humus and charcoal. Soil always contains a small amount of charcoal that is the result of fires. It is very stable and will last for hundreds of years, but it does not impact soil or plant growth to any great extent.

What is the definition of *organic matter*? That is not entirely clear. Some sources include all three of the above-mentioned forms of carbon. Others include only the dead and the very dead, and still others include everything that contains carbon, including large molecules like proteins and carbohydrates. The actual definition is not that critical, unless you are testing the organic matter level in soil, and then it is good to know what is being measured and reported.

Soil samples usually exclude larger living organisms, such as earthworms, and larger plant roots. The lab then tests for any remaining organic material, which includes living organisms and free organic molecules.

Soil can contain amounts of organic matter varying from 0% to 90%. Mineral soils generally have values in the range of 3% to 6%, based on weight, and the ideal level of organic matter depends on the soil texture. Sandy soil with 2% OM is very good, and getting values higher than 2% can be difficult. However, clay soils with 2% OM are considered low, and 5% is considered good. All of these numbers refer to the OM in topsoil. Soils with more than 20% OM are considered organic soil and are found in places like peat bogs.

The percentage value of organic matter can be confusing. Labs will normally report it on a weight basis, as I have done in this book. Online sources, especially ones talking about mixing soils, usually measure by volume and therefore report much higher values. A 10%, by volume, is about the same as 5% by weight. Unfortunately, most sources don't indicate if the value they are reporting is by weight or volume.

Organic matter is a small component of most soil, but it has a great affect on almost all soil properties, and increasing levels in marginal soil will have the following physical effects:
- Increased aggregation
- Improved water infiltration that reduces runoff
- Increased aeration
- Increased water-holding capacity (holds 6 times its weight in water)
- Improved tilth of clay soils
- Reduced surface crusting
- Improved pore spaces

Increased levels of OM will also have chemical effects.
- Increased CEC (cation exchange capacity) increases cation levels
- Increased availability of nitrogen, boron, molybdenum, phosphorus, and sulfur
- Increased microbial activity
- Increased microbial diversity

Adding OM to soil can also have some short-term negative effects. It can tie up nitrogen if the source has a high C:N ratio. Some plant residue can be toxic to other plants. For example, dead quackgrass roots can slow down the growth of crop plants.

Adding fresh OM is better than adding decomposed material because it provides more nutrients and has a higher level of binding agents to form aggregates. When gardeners consider adding nutrients to the garden, they think in terms of compost and manure. Although these are a prime nutrient source, their real value is in providing food for the many microbes that reproduce quickly and are short-lived. This constant turnover of life has the effect of multiplying the amount and value of organic matter being added to soil.

Organic matter is also a major sink for carbon and typically consists of about 60% carbon. Plants capture CO_2 from the air and turn it into relatively stable carbon. When plants die, decomposition starts releasing CO_2 back into the air. Traditional agricultural practises speed up this process and add to the global CO_2 problem. As good stewards of the Earth, we need to maintain and even increase the carbon level in soil.

The amount of OM in soil is a balancing act. Loss occurs due to decomposition and erosion. It is our goal to garden in such a way that the rate of loss is reduced and to continually add fresh material to compensate for the losses.

Decomposition: Converting Dead Things into Humus

It is critical for you to understand the process of organic matter decomposition in order to comprehend how various gardening practices affect OM in soil. It all starts with living organisms, everything from microbes to earthworms to plants. At some point, these die, producing what is affectionately called fresh organic matter.

Animals and plants consist of one or more cells. The outside of the cell, the cell membrane, varies greatly, but its main function is to contain the cell contents. These consist of a very wide range of compounds including simple nutrients like nitrates and phosphates and more complex ones like sugars, amino acids, right up to large molecules like proteins, fats, and carbohydrates. At death, these cells stay intact.

Not much happens immediately after death. The fall leaves just sit there, and in colder climates, they are still there in spring. The first step in decomposition is that arthropods, earthworms, and fungi start ripping the leaves into smaller pieces. At this point, cell walls start to break apart, releasing their contents.

You can think of the cell contents as a soup containing thousands of chemicals in varying sizes. All of these are suddenly available for living organisms to use. Plant roots can absorb the simple nutrients, and microbes absorb the sugars. Most of the larger molecules can't be used and must wait for further decomposition. The cell walls are made from tougher material that requires special microbes. Fungi are one of the few organisms that can decompose lignin and cellulose.

This mineralization process is much slower than most people realize. We talk about "finished compost," which looks like soil, but it is mostly undecomposed material. It will keep decomposing for several years, depending on the source of the material and environmental conditions, but you can use a rough figure of five years before completion.

The process is slower in cooler, drier climates, and clay protects OM from decay. Drainage also has an effect. Because wet areas have lower levels of oxygen that is essential for decay, decomposition happens more slowly. This explains the existence of bogs, which contain lots of partially decomposed OM.

During the five-year mineralization period, nutrients are slowly released. Some is used by plants and other organisms, but much of it becomes attached to soil particles, where it is stored for future use. This slow and continuous release of nutrients is the major reason organic matter is so good for plants.

The only thing left at the end of the process is humus, a mixture of carbon, oxygen, and hydrogen. It has been described as being a very stable complex molecule that is resistant to further decay. Experts estimate the age of some humus to be 1,000 years old.

Since plants are able to get their carbon, oxygen, and hydrogen from the air and water, humus is not of any nutritional value to plants. However, it plays a critical role in conditioning soil. Many of the benefits ascribed to OM are actually due to humus.

Truth About Humus

In the previous section, I described humus as it has been described for many years: a very stable component of soil. The truth of the matter is that this vision of humus is almost certainly wrong.

Some clarification of the term *humus* is helpful since many people use it incorrectly. Gardeners talk about humusy soil and buying humus in bags. In this context, the term is used to describe an ideal black crumbly soil, and suppliers of soil tend to label their products with the word "humus" to make buyers believe their product is superior. The reality is that you can't buy humus in bags, and none of these products are humus. Most are compost or a mixture of soil and compost.

Humus has been studied for over 200 years by treating soil with a strong alkaline solution. This produces a black material that is called humus. Much effort has gone into defining the chemical structure of this material, but even after all this time, the structure is still unknown. The other big problem with humus is that scientists can't find it in soil. They can only see it, and test it, after extracting it.

One of the confusing aspects about humus is that, during the decomposition process, organisms continuously convert large molecules into smaller and smaller ones. You would expect the end product to be small molecules, and yet at the end of the process, they have created humus that consists of much larger molecules than most of the ones they started with. Humus is also very stable, and even microbes can't decompose it, even though microbes can decompose just about everything that is carbon based. No one has been able to explain these observations.

The problem with studies done with extracted humus is that a strong alkaline solution completely changes the material during the extraction process. It is kind of like putting a variety of vegetables through a food processor and then trying to identify the original vegetables by looking at the resulting smoothie. Given all of the advanced chemical techniques they now have, scientists have come to the conclusion that humus does not really exist in soil.

The decomposition process described in the previous section discussed this as an isolated, one-way process. Large molecules are continually broken down into smaller ones. What the discussion ignored is a parallel process where microbes immobilize the small molecules into large ones. In effect, microbes slow down the decomposition process by making it start all over again, but this time the fresh organic matter are the dead microbes themselves.

Decomposition is really a two-way process. Mineralization and immobilization are taking place simultaneously.

Soil Myth: Humic Substances Exist

The extracted humus can be further treated with chemicals to produce a number of other compounds, collectively called *humic substances*. The following three are the more commonly discussed ones:

- Fulvic acid has smaller molecules that are soluble in acids and alkali and are degraded by microbes.
- Humic acid consists of medium-sized molecules and is soluble in alkali but not acid. It is resistant to microbial attack.
- Humin has the largest molecules and is darkest in color, insoluble in acid and alkali, and unaffected by microbes.

Since humus is perceived as having great value for soil and plants, it follows that these special fractions of humus must also have value; this has resulted in many companies selling these items. Keep in mind that these are not natural compounds and are not found in soil. They should not be an acceptable amendment for organic gardening. There is no scientific evidence that these compounds have additional benefits over and above natural organic matter.

In this scenario, humus is not very stable; instead, it is in continual flux. At any given time, a wide range of molecules are available. This description is much more consistent with current scientific data than the one that describes stable humus.

Why have I given you different descriptions of humus? The one in this section provides a true picture of current scientific thinking, but that could change again. After all, it's been discussed for 200 years. The definition in the previous section is the one you will find in most gardening literature, and it does have some educational merit. The concept of stable humus is certainly easier to understand than a system in constant flux.

The Scientific Process

There is significant scepticism these days about science and the scientific process. People expect science to always be right, but they fail to understand that it is an iterative, self-correcting process. A scientist's job is to develop ideas and then test them. Many times, the ideas are wrong, and it can take years to correct them. That is all part of the scientific process.

Science has been wrong about humus for 200 years, but even as early as 1888, some scientists were sceptical about its existence. Then again 50 years ago, there was significant movement to redefine it. There is now more support for the new idea of humus, but it may change again as our understanding of soil and our chemical techniques improve.

The important point of all this is that science research is a self-correcting process, and over time, science does reach the right conclusions. We have just observed, for the first time, the gravitational waves predicted by Albert Einstein in 1916. Science may be slow at times, but it is much better than the alternatives.

Too Much Organic Matter

The above description paints a rosy picture for organic matter, clearly showing that it does all kinds of great things for soil and plants. But can you have too much of a great thing? The answer is yes. Too much organic matter can be toxic to plants.

Adding an excess of organic matter to a new garden or to an older agricultural field is usually not a problem. Such soil almost certainly has a low level of OM, and adding a bit too much can only help the garden. The problem occurs when this is done year after year, especially in small gardens like raised beds. People are making new raised beds with 30% compost and adding more each year.

The initial effect of adding fresh OM is an explosion of microbial growth, and they need nitrogen to start decomposing. This reduces the nitrogen level in soil and starves plants. A balance is reached over time, and nitrogen is again made available to plants.

Remember that OM takes several years to decompose fully. If more OM is added each year, there is a steady accumulation of undigested OM in the soil. After a few years, the amount is so large that released nutrients can reach toxic levels.

The amount of NPK used by plants varies by species, but an average value is around 7-1-6. They use about 7 times as much nitrogen as phosphorus. Most commercial manure-based composts have an NPK of around 1-1-1. If you add enough of this to supply the needed nitrogen, you will add 7 times too much phosphorus. Since phosphorus does not travel through soil quickly and is converted into a stable form, it accumulates in soil. Excess nitrogen easily leaches away. Repeat this annually, and you can see why soils quickly reach high levels of phosphorus.

A high phosphorus level makes it more difficult for plants to take up manganese and iron, resulting in an effective deficiency of these nutrients in plants. This shows up as interveinal chlorosis of the leaves. Some people try to solve this problem by adding more iron, but if the problem is caused by too much phosphorus, adding more iron won't help. High phosphorus levels are also toxic to mycorrhizal fungi, which then don't associate with plants that then need to form more roots to find their own phosphorus. This can reduce flowering and fruiting.

How much OM is too much? That is difficult to determine without a soil test. The amount certainly depends on the soil texture, cultivation practices, and the environment. It also depends on the source since plant-based material has a relatively smaller amount of phosphorus than manure-based material. The negative effects of too much OM are also cumulative and slow to show themselves.

In recent years, soil testing of small high-intensity organic farms has reported very high levels of phosphorus, along with numerous plant growth issues. One of the best ways to detect an OM problem is to monitor phosphorus levels. If these get high, stop adding organic matter.

Compost

The term *compost* is used in various ways. In the UK, it is equivalent to the American term *potting soil*. It is also used to describe any type of OM. In this book, it describes partially decomposed OM, the material you harvest from a compost pile after about six months.

The properties of compost depend very much on the material and method used to make it. The nutrients in compost are relatively low compared to synthetic fertilizers. A study looking at home composts found an average NPK of 3-1-1.5. Many commercial compost products have an NPK of 1-1-1.

The section on decomposition describes in general terms the process of making compost. For the most part, it is a microbial process that slowly converts large physical pieces of material into smaller and smaller components, until you are left with basic nutrients, humus, CO_2, and water.

The composting process is much more complex than this. As the process proceeds, the microbial populations are continually changing. Initially the pile of material is cool and has very little free nitrogen. Fungi are the main active microbes at this point.

As material is broken down and the content of cells is released, nitrogen becomes more available. This encourages the growth of bacteria that enjoy a cool environment. Their activity starts heating up the pile, and as temperatures change, completely new populations of microbes develop.

Soil Myth: Compost Is Acidic

The pH of compost depends on the input ingredients. In the initial stages, organic acids are formed, which make the compost pile more acidic—the pH drops. In these conditions, fungi grow better than bacteria, and they take over the pile and start to decompose the lignin and cellulose in plants. As a result, the pH rises and bacteria become more populous. Finished compost is usually alkaline, with a pH between 7 and 8; for example, yard debris 7.7, mixed manure 7.9, and leaf litter 7.2. The ones below 7 include horse manure at 6.4 and bark compost at 5.4. Most homemade garden compost has a pH between 7.0 and 7.5.

Eventually, nitrogen becomes scarce again and microbes start dying off. The pile starts to cool, and soon you have what gardeners call finished compost. This is then added to the garden, where the composting process continues for several years.

A critical factor that controls this whole process is the carbon-to-nitrogen ratio (C:N ratio).

C:N Ratio

The common advice for making compost is that you should use the correct ratio of browns and greens. This an attempt to get the right C:N ratio so that composting takes place quickly and at a high temperature.

Growing organisms need a certain amount of carbon and nitrogen to grow. The ideal food source for them has a ratio of carbon to nitrogen of around 30:1. Most compost piles have extra carbon and not enough nitrogen. When nitrogen is too low, the metabolism of all life forms slows down, which also slows down the composting process. Slow composting prevents the pile from getting hot enough to kill parasites and weed seeds. High nitrogen levels do not impact microbial growth as much since the excess is usually lost as it's converted to atmospheric nitrogen.

Soil Myth: Ratio of Green and Browns

A common recommendation is to mix browns and greens in a ratio of 30 to 1, but this is completely wrong. The idea of a brown-to-green ratio might be easy for gardeners to understand, but it leads to a misunderstanding of the process. It is actually the carbon-to-nitrogen ratio that should be 30:1, and it would be better to use terms like high-nitrogen ingredient and low-nitrogen ingredient in place of greens and browns.

In the simplest form, the terms are quite descriptive. Browns are any plant material that is brown, including fall leaves, dried grass, wood products, paper, and straw. Greens are—you guessed it—green, including fresh grass clippings, weeds, plant clippings, and most kitchen scraps.

Brown and green materials contain carbon and nitrogen, and each ingredient will have its own C:N ratio. For example, horse manure is about 25:1. Fall leaves have a ratio of about 50:1, depending on age and type of leaves. Manure is brown, but it is considered to be a green since it has a high relative amount of nitrogen. Fall leaves are brown, but if they fall in a green state, they are greens. It all gets very confusing.

How do you mix manure and fall leaves? If you follow the common rule of 30:1, brown to green, you would add 30 parts leaves to 1 part manure, but this results in a C:N ratio of 49:1. If you use the C:N ratio and a target of 30:1, you would add 4 parts manure to 1 part leaves.

The ratio of material in a compost pile is even more complicated. The reported C:N ratios are weight-based, and most gardeners add material on a volume basis. Water is a big part of the weight, especially for green material. Although there are lists of ratios for ingredients, most gardeners don't know the values for their input ingredients. The process is further complicated by the fact that most gardeners have more browns than greens, unless they bring in some manure or have a cow in the backyard.

C:N Ratio of Various Sources of Organic Matter.

Material	C:N Ratio
Alfalfa	12
Cardboard	350
Coffee grounds	20
Corn stalks	75
Food waste	12–30
Grass, fresh cut	15
Leaves	60
Manure	5–25
Newspaper	170
Pine needles	80
Straw/hay	75
Weeds	30
Wood ash	25
Wood chips	400

Some people recommend adding compost boosters or compost starters, but these are unnecessary. They generally contain some microbes and possibly some extra nitrogen. The microbes are not needed since all of the plant material you put into the compost pile is covered with microbes. Nitrogen may help if the C:N ratio is too high, but fertilizer, especially urea, is a cheaper alternative.

The reality is that if you simply add your ingredients to the compost pile when you get them, and you turn the pile, you will make compost. It might be a slower process, but you can't stop decomposition.

Hot Composting

The process described above is hot composting. It has many variations, but it generally consists of making a big pile of material and trying to get the C:N ratio correct. It is kept moist and turned periodically to allow air to get into the pile, which will heat up, which in turn speeds up the decomposition process.

In a hot climate, compost is ready to use in a few months. Commercial operations, which control all aspects of the process, can make compost very quickly.

Cold Composting

The problem with many smaller operations is that they don't have material with the right C:N ratio available all at the same time; much of it is available in fall and has a low nitrogen level. An alternative to hot composting is cold composting. Pile up the material and let it sit.

The lack of nitrogen means that the process is much slower, usually more fungal than bacterial, and the pile does not heat up. The resulting material is commonly called leaf mold, which is partially decomposed leaves and fungal hyphae. It is still good compost, containing essentially the same nutrients that are produced from hot composting, but it may have more live seeds and pathogens.

Cold composting also happens in smaller piles and fancy rotating drums that can't generate enough heat.

Vermicomposting

Vermicomposting uses a bin, organic material, and worms. The worms eat the organic material and produce worm castings, i.e., worm poop, that are used to amend soil.

Although it is called composting, it is not really a form of this process. It should be called vermidigestion, not vermicomposting. The result is a form of degraded organic matter, but it is not clear how similar this is to compost.

The process is normally done in small containers in home basements, but it is also suitable for larger operations. The bins can be any size, and many could be used. Long bins are common in commercial operations, where fresh organic matter and worms are added at one end. New OM is added next to the first pile, and worms naturally migrate to it, leaving behind their casts. In this way, worms move along the bin until they reach the end, and then the process is started all over again. The worms need to be kept warm, so in cold climates they must be kept indoors.

Bokashi Composting

Bokashi composting in North America is usually described as a small in-home operation. You keep a bucket under the sink and add

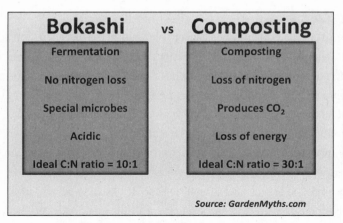

Comparison of bokashi and composting.

kitchen scraps to it. Over time, you get something called bokashi ferment that can then be added to the garden.

In other parts of the world, bokashi composting is done on a much larger scale, using piles that are not unlike those used for hot composting. Input ingredients usually consist of some type of plant refuse and manure.

Although called bokashi composting, it's really a fermentation process that is started with a special mixture of microbes, not unlike the process of making beer and wine. The inputs are then fermented in an anaerobic environment.

Decomposition does not really happen until the material is added to soil in an aerobic environment. The concept is based on the idea that the fermented material decomposes faster than unfermented material, but it's unclear whether this is really true.

Field Composting

Field composting includes any process in which organic matter is left in the field for nature to do its own thing. No-till is one form where plant refuse is left right in the field to decompose. Cover cropping is a similar method, which produces a lot of refuse.

I use this method in my own garden and call it the cut-and-drop method. Any cuttings and deadheads, including most weeds, are dropped right in the garden. I might toss it behind a bigger plant so it is not visible, but it all stays right where it grew.

Composting: Which Method Is Best?

There is much discussion about which composting method is best, but in reality they are all about the same. One might be faster than another, and that may or may not be a benefit. Remember that processes in nature are slow.

What a composting process produces has more to do with the input ingredients. If you take a handful of manure, you have the nutrients found in a handful of manure. None of the methods described above will increase the amount of nutrients.

Composting can have an effect on the potassium and nitrogen content of the finished product. As cells break apart, potassium is easily lost because of leaching, since it is not tied up in large molecules. Microbes also convert nitrogen to atmospheric nitrogen, and so it can be lost. For both these reasons, it is usually best to partially compost material and let most of it happen in the field.

The best method of composting is the one that you do and continue to do because you like doing it. Any form of composting is better than taking yard waste to the curb.

Soil Myth: Compost Starters and Accelerators

Compost starters are products to help start the composting process and usually contain so-called vital organisms. These products are not required. All of the microbes you need for composting are already covering the organic matter you are adding to the pile. If you really feel the need to add more, toss in a handful of soil.

Compost accelerators can speed up composting by adding some nitrogen if the C:N ratio is too high. You don't need special products for this; just add some high-nitrogen fertilizer. It does not require much, and if you add too much, the pile will smell like urea (urine).

Compost NPK Values

Compost tends to have low levels of nutrients, with nitrogen and phosphorus being 1% to 3% and potassium in the order of 1%. A common average for commercial compost is 1-1-1.

Nitrogen in compost is available as nitrogen ions (5%), which are available immediately, and large organic molecules (95%), which need to decompose before nitrogen can be used by plants. Almost none of the nitrogen is available immediately for the plants to use, and only 10% to 30% is released in the first year. This is why synthetic nitrogen is a better option for plants that need nitrogen quickly, like a new vegetable garden. Compost is a long-term slow-release form of nitrogen.

Compost may contain as much phosphorus as nitrogen; almost all of it is tied up in large molecules, so it too is in a slow-release form.

Potassium does not get incorporated in large molecules and is always in an ion form ready to be used by plants. It is also very soluble in water, and rain or ambitious hose watering can easily leach potassium out of the compost pile before it is finished. This is one reason some people cover their compost bins.

Chelation

Several of the important micronutrients, such as iron, copper, zinc, magnesium, and manganese, exist as metals when they are in their elemental form. These metals combine with oxygen, forming a type of rust very similar to the process of iron rusting.

The same reaction takes place in soil. Once these metal ions oxidize (i.e., rust), they become insoluble and can no longer be used by plants. These cations can also interact with other anions, forming insoluble salts. All of these chemical reactions effectively remove nutrients from the soil solution. The soil may still contain lots of nutrients, but they are in a form plants can't use.

Chelates are organic molecules that pick up and hold the metal ions so that they don't react with oxygen or other ions. Even though chelates hold on to the metal ions, plant roots still have access to them, thereby ensuring that plants have access to a higher level of nutrients. Think of chelates as small taxis that drive nutrients around until they bump into a plant root. They then let their passenger out and go find another nutrient ion.

Plants even use chelates internally to move metal ions from roots to leaves, ensuring that they don't react before they get to their destination. Chelates are clearly important for plant growth and provide the following benefits:

- Keep nutrient ions soluble
- Prevent precipitation
- Overcome the effects of pH on nutrient ions
- Reduce toxicity of metal ions
- Reduce leaching
- Suppress plant pathogens

There are commercial sources of chelates with active ingredients like EDTA, DTPA, and EDDHA. These can be added to soil to provide the above benefits. You can also buy chelated fertilizer, in which the micronutrients are attached to chelate molecules during the manufacturing process. When the fertilizer is added to soil, the ions are immediately protected from oxygen and reactions with other ions.

The natural source of chelates is organic matter, which is a principal reason it is necessary for soil. In effect, OM holds on to nutrients until plant roots find them, in part explaining why soils with a higher OM level are more nutritious.

CHAPTER 8

Rhizosphere

The rhizosphere is a very special and unique area in soil. Microbe populations can be a thousand times higher here, and it is full of chemicals that are excreted by both plants and microbes. It also contains a very high level of fresh organic matter, and the pH can be quite different from elsewhere.

This unusual place is defined as being the 2- to 3-mm area around roots, especially near their tips. The properties of the rhizosphere are concentrated next to the roots, but they gradually thin out as you move about 1 cm away from the root.

Because of the high population of living organisms, this area has lower pH, higher organic matter, higher CO_2, lower nutrient content, fewer contaminants, and lower water resources than the rest of the soil because there is such a high competition for resources.

Although this small area is critical to plant growth, it's not well understood by gardeners or farmers. Even scientists are just now starting to understand it.

In many ways, soil resembles our rural and urban areas. Rural area accounts for most of the soil space, but it accommodates only few people. As you move toward the city, the population increases and things get more congested. The population density is highest right downtown where the roots grow. Most of human activity takes place downtown.

Root Exudates

Roots release all kinds of chemicals, which as a group are referred to as *root exudates*. Scientific research has shown that they are produced for various reasons:

- Restricting the growth of competing roots
- Attracting microbes in order to form symbiotic relationships
- Changing the chemical and physical properties of the soil and soil solution
- Making nutrients more available

The exudates include sugars, carbohydrates, and proteins that form a perfect food source for microbes. Sugars and carbohydrates provide the carbon, and proteins provide the nitrogen, the perfect buffet for living organisms.

When a root enters a new area of soil, the first to join the party are the bacteria and fungi. Bacteria in particular show explosive growth as they feed on the exudates. This is quickly followed by their predators: nematodes and protozoa. As the microbe populations increase, they each contribute their own excrements, adding to the nutrient pool. The water layer around a root becomes full of thousands of various chemicals.

Plant roots also produce amino acids, vitamins, organic acids, nucleotides, flavonoids, enzymes, glycosides, auxins, and saponins. Some of these attract specific organisms that plants want to have nearby, and other chemicals can be toxic to certain species. What is amazing about this is that, through the use of exudates, plants have some control over the type and number of microbes that inhabit the rhizosphere. They even change the populations based on seasons and environmental conditions.

The exact nature of the exudates depends on many factors, including genetics, age of the plant, light, water, temperature, mineral deficiencies, and even reduction of leaf surfaces due to pests and diseases. For example, at 98°F (37°C), the amount of exudates increases for the warm-growing kidney beans but decreases for the cool-growing peas. The peas are done for the year, so there is no point in producing more exudates. In wheat, seedlings and mature plants can have six times fewer exudates than young growing plants in the tillering stage.

Herbicides, antibiotics, and fertilizer sprayed onto plants also affect the exudates produced, and we know very little about these effects. Producing these chemicals is a major drain on plant food resources. It is estimated that some plants excrete 50% of the fixed carbon from photosynthesis, through the roots. Annuals excrete 40% of their photosynthates; for most plants, the typical value is around 30%.

Producing these exudates is very costly to the plants, and so they must get significant benefits from it, and they do. The large number of microbes results in an abundance of dead microbes, due to both short life spans and predation. This produces high levels of nutrient ions right next to the roots. This is much more efficient than trying to find and extract nutrients out of soil.

Most of the activity in the rhizosphere occurs near the tip of the root. This is the point where most of the water and nutrients are absorbed by the plant, so it makes sense to also produce the exudates here. Root caps and root hairs are also a vital component of the rhizosphere since they are short lived. Root caps sheath off up to 10,000 cells each day, and root hairs live only a couple of weeks. Combined, these two sources add a significant amount of fresh organic matter, a perfect food source for microbes.

Soil Enzymes

Enzymes are special protein molecules that are responsible for most chemical reactions in a cell. Making large carbohydrates out of simple sugars is done with enzymes, as is breaking large molecules into smaller ones. The soil solution contains many enzymes. Microbes excrete them, and dead cell walls break apart, releasing even more. This soup of enzymes can attack microbes, digest organic matter, decompose other enzymes, and be absorbed by other microbes and by clay particles. An enzyme called *urease* has been found in permanently frozen soil that is thousands of years old. This concentration of enzymes is perfect for decomposing large molecules into smaller ones so that plants and microbes can use them.

Plants are quite sneaky in driving this whole process. They produce exudates that attract a wide range of organisms, which in turn produce the majority of enzymes, which degrade organic matter and produce nutrients to feed the plants.

Effect of Desiccation

All of the organisms in the rhizosphere depend heavily on the presence of moisture. Due to the high density of organisms, water is used up quickly. As this happens, more water flows in from surrounding soil particles, bringing with it nutrient ions.

At some point, the water level in soil drops too much, and the rhizosphere starts to dry up, causing a steady decrease in microbe population and root hairs to die off. All of this decaying organic matter results in an increase of ammonium that is lost to the air, reducing the nitrogen available to plants. This drying process changes the chemistry and biology of the rhizosphere drastically, making it hard for plants to obtain the needed water and nutrients.

When water returns, there is an explosion of microbial growth that feasts on all of the dead organic matter. This drastic increase in growth makes nitrogen scarce and levels drop. Plants are suddenly deprived of nitrogen.

Clearly, it is best to keep plants watered so that these shortages do not occur.

Soil pH Levels

Bacteria and fungi are both important to the rhizosphere, but their presence depends on pH levels. Fungi are more dominant in acidic conditions because bacteria have difficulty growing there. Fungi can grow in a pH range of 3.5 to 8.5, while bacteria have difficulty living in soil below 6.5.

As part of the exudates, plants can also produce acids to modify the pH to suit their needs. The soil solution around a root tip can be as much as 2 pH units different from other areas of the soil.

Nitrogen is absorbed by roots as either the ammonium ion or

nitrate ion. Plants need to maintain a neutral charge. So when they absorb a positively charged ammonium ion, they release a positive hydrogen ion. which makes the rhizosphere more acidic. When the negative nitrate ion is absorbed, the root releases a negative hydroxyl ion (OH-), which increases the pH. The ionic form of the fertilizer you use can significantly affects the pH of the rhizosphere. The nitrogen-fixing nodules of legumes result in the plant releasing hydrogen ions, making the rhizosphere more acidic.

The large population of organisms in the rhizosphere all absorb oxygen and respire carbon dioxide just like animals. Even roots excrete CO_2 as they carry out respiration. When CO_2 dissolves in water, it produces carbonic acid, lowering the pH.

The net effect of all of these chemical reactions is that the pH of the rhizosphere is usually much more acidic than the rest of the soil and helps explain why plants can grow in alkaline soil. The roots are actually growing in acidic conditions.

The fixation on soil pH by many gardeners may be unfounded, at least to some extent. Plants may control their pH environment much better than we think.

Dynamic Microbe Population

The population of microbes in the rhizosphere is complex and varied. Many factors affect the type and number of microbes. The CO_2 level in air is 0.04%, and in soil it normally ranges from 0.3% to 5%. Due to the high number of microbes, the level in the rhizosphere can be as much as 20%. As the level rises, microbes are affected; some prosper and others perish.

The type of compounds produced during decomposition of organic matter also changes depending on the materials. Many browns, like fallen leaves, produce alcohols and aldehydes, which can be toxic to certain bacteria. Decomposition of wood yields chemicals that can affect fungi. Even in death, plants affect the microbes in soil.

The number of microbes in the rhizosphere is high, but the diversity is much lower than in the rest of the soil. Because they multiply faster in the rhizosphere, there is also a higher turnover, and populations can change more quickly. Much of this is controlled by the plants.

There are many beneficial interactions between microbes and plants:

- Specific exudates attract mycorrhizal fungi to the roots.
- Seeds produce different compounds before germination than after germination in an effort to manipulate the microbe populations in their favor.
- Legumes produce flavonoids to attract rhizobia so that they can form nitrogen-fixing nodules.
- Special bacteria called plant-growth–promoting bacteria colonize the rhizosphere and produce plant hormones (auxins and cytokinins) that increase plant growth and improve mineral uptake.
- Microbes can also produce chelates that make nutrients more available.
- Inoculation of potatoes with bacteria has shown a 17% increase in yield, probably due to the antibiotcs they produce, which help control pathogens.
- Other bacteria release enzymes that inhibit the development of nematode eggs, decreasing root damage.
- Azospirillum are bacteria that live around the roots of grasses and fix nitrogen.

Science is just now starting to discover and understand these associations, but it is clear that the microbe life in the rhizosphere is critical for plant growth.

Allelochemicals

Allelopathy is a biological phenomenon in which an organism produces chemicals that affect the growth and survival of another organism. Allelochemicals can have beneficial or detrimental consequences. In the strict definition, this applies to any type of organism, but the term is commonly used by gardeners to describe interactions between plants, and most of the discussion centers around detrimental effects.

The most well-known case of allelopathy is the effect that walnut trees have on the growth of other plants. Many believe that nothing grows under a walnut tree because it excretes a chemical called juglone. In fact, the various parts of a walnut tree don't contain any juglone. However, they do have another compound that turns into juglone on exposure to air, but this quickly converts to other compounds. Juglone will prevent the growth of some seedlings, notably tomatoes and other nightshades, but most of the studies showing this have been done in the lab and not the real world. Nobody has been able to demonstrate that plant roots actually take up juglone from the soil. If they don't absorb it, how can it affect plant growth? It is true that some plants don't grow as well under a walnut tree, but most mature ones grow just fine.

Other examples of allelopathy include garlic mustard (*Alliaria petiolate*), which inhibits the growth of other plants and mycorrhizal fungi. Mexican fireweed (*Bassia scoparia*), an invasive weed in central North America, reduces the growth of spring wheat.

Companion planting is usually seen as a way for using one plant to help the growth of another, but there are also negative effects, and companion planting rules also list combinations that should not be placed together. I'll discuss the validity of these ideas elsewhere, but if they are true, then it is almost certain that many of the negative plant combinations are a result of allopathy.

Many other examples of plant-to-plant allelopathy exist, but most of their details are not well understood. The chemical effects of allelopathy reach beyond the rhizosphere but weaken as one gets farther away from plant roots.

Plants Are in Control

We think of plants as being passive about getting their nutrients. Plants do grow roots in search of nutrients, but this growth is fairly random because the plant does not know where the nutrients are until a root stumbles onto them. This helpless view of plants is misleading. Plants play a major role in bringing nutrients to the roots.

As discussed above, plants are actively herding and attracting the right microbe food sources to live next to their roots, but they even take an active role in extracting nutrients from soil. Acidic exudates change the pH of the soil solution around the roots, which helps dissolve insoluble minerals in the soil, producing useable nutrients. In effect, the acids are dissolving the rocks around the roots.

These changes in pH increase the availability of micronutrients like zinc, calcium, and magnesium, but they are most important in making phosphorus more available. The form of phosphorus that plants use is phosphate (PO_4), which develops insoluble compounds with calcium, aluminum, and iron, hiding it from roots. But at a more acidic pH, the phosphate is released so that plants can use it.

Iron deficiencies cause some plants, like grasses, to exude chelates that combine with iron molecules. The flow of water toward the roots then brings this chelated iron to the surface of the root, where it is able to strip out the iron molecule and absorb it. The remaining chelate molecule then returns to the soil solution looking for another iron molecule.

The production of all of these exudates is very much under the control of the plants. If iron levels are low, they produce chelates. If phosphorus levels are low, they produce more acids. As nutrient levels rise, they produce fewer sugars because they no longer need a big microbe herd. Why feed them and encourage their growth when the plant has enough food?

Plants manipulate the rhizosphere and its populations to benefit themselves. It is important to understand that the previous sentence does not imply any sort of knowledge, thinking, planning, or intelligence on the part of plants. All of this is controlled by basic chemical reactions, many of which are controlled by enzymes that have the capability to change their activity based on the presence of certain triggers. For example, at high iron levels, an enzyme may be present but inactive. When levels drop, it becomes more active, which results in the production of a special chelating exudate, making it easier to absorb iron molecules. As iron levels increase, it again shuts down activity until needed. This all happens through the magic of chemistry—not intelligence.

Solving Soil Problems

CHAPTER 9

Identifying Soil Problems

Many people garden by following the recommendations of others. For example, they lime their soil annually because some gardening guru said it should be done. The problem with this approach is that you may be wasting natural resources and your time, or worse, you could be harming your soil's health.

Before you do anything in the garden, try to understand why you are doing it. What effect will it have? What problem are you solving? If you are not solving a problem, then don't do it. Many standard gardening practices simply are not required.

In this chapter, I will look at specific soil problems. As you read through each section, ask yourself if they apply to you. Most soil will not have all of these problems, and there is no point in solving the ones you don't have. However, learning about all the problems will give you a much better insight into the workings below your feet.

Why Do We Fertilize?

Most gardeners fertilize because someone said they should do it, or some online source suggested it. They do it in the mistaken belief that plants need to be fed; this is completely wrong. You do not fertilize plants.

You fertilize to replace missing nutrients in soil.

You may have a hard time believing this since it goes against everything you have read, including so-called expert advice from top gardening gurus, specialty plant societies, and fertilizer manufacturers.

The idea that we must feed the plant seems to make perfect sense, but it ignores a critical point. They get their food from the soil solution. You don't add fertilizer to plants—you add it to soil. It is best explained by a simple example. Assume your soil is low on phosphorous, but it has plenty of nitrogen and potassium. The neighbor across the street has soil that is high in phosphorus and potassium but is missing nitrogen. Should you both use the same fertilizer to grow tomatoes? Clearly the answer is no. You need to add phosphorus, and your neighbor needs to add nitrogen. The fertilizer you add to the garden should supply the nutrients missing from your soil.

Still not convinced? Go to Google and use the image results to search for tomato fertilizer. You will see a page full of specialty tomato fertilizers, all having a different NPK formula. How can each product be the perfect formulation and yet be different from all the others? At best only one is correct, but that even that is unlikely. In reality, there is no such thing as tomato fertilizer, except in the minds of marketing departments and, unfortunately, their customers.

Plants absorb the nutrients they need from the soil. If grass needs more nitrogen, it takes more nitrogen from the soil than a plant that needs less. If a plant is ready to bloom and it needs more phosphorus, it takes more from the soil. Your job is to keep nutrient levels high enough so that plants never run short.

How do you know if your nutrients are low? You have to test your soil.

Soil Testing

Soil testing can provide information about the texture, pH, and levels of nutrients and organic matter. With the exception of measuring texture (see section below), the best approach is to have a commercial lab perform these tests.

In the US, most states have extension offices, usually associated with a university, that will do soil testing at very reasonable rates. In Canada, tests can be done through private labs and the University of Guelph. In Europe and Australia, testing is normally done at a private lab.

A basic soil test will normally measure pH, phosphorus, potassium, calcium, and magnesium. You can also request extended tests for other nutrients, organic matter, pollutants, and heavy metals.

Many soils do not have a micronutrient deficiency, so unless you suspect a problem, it is probably not worth testing for these. Talk to your local lab about local deficiencies. They will have a good idea what should be tested.

You might have noticed that nitrogen is not mentioned in the above list. This is because nitrogen levels change very quickly at warmer temperatures. By the time you get samples to the lab and see results, the values have changed. So there is not much point testing for it, and the test is fairly expensive. Larger agriculture operations will take soil samples, freeze them, and keep them frozen until tested, so that values don't change in the sample.

Nitrogen is the mostly likely nutrient that is in short supply, and it is the one that is not tested. Most labs estimate the amount of nitrogen you should add, based on other nutrients, the organic matter result, and local experience.

Should You Get a Soil Test?

The advice from everyone is to get a soil test, but I am going to break with tradition. Soil tests have limitations, and they are not always worth getting. If you operate a large agricultural business, or even a smaller market garden and yield is important to your business, you should get a soil test. If you are a homeowner looking to grow landscape plants and maybe even have a small garden for vegetables, it is probably not worth the cost.

Besides the size of your operation, ask yourself this question. What are you going to do with the results? Are you going to add the suggested fertilizers so that you only add the missing nutrients? If so, the test makes more sense. On the other hand, if you will be using manure or compost, which provides all of the nutrients; skip the test. This advice also applies to gardeners who will look at the results and then use a 10-10-10 fertilizer because it was on sale.

A soil test can give you some quick insights into a new garden, but if you have been gardening in a place for a while or moved into a home with established gardens where things are growing well, you probably don't need a soil test. If plants are not growing well, you could have a nutrient deficiency, or a nutrient might have reached toxic levels. In this case a soil test is a good idea.

You probably don't need to fertilize if some of the following apply:
- You mulch with an organic material that breaks down over time.
- You add compost on a yearly basis.
- You return spent plant material back to the garden instead of sending it to landfill.
- Your plants are mostly growing well. Not all plants do well in all gardens; accept this fact. Adding fertilizer is rarely the answer.

What Do Tests Really Measure?

When you get a soil test done for nitrogen, phosphorus, and potassium, you might think that they measure the total amount of these nutrients, but that is not quite correct. Nitrogen is usually not measured for soil, but if it is, the usual test is a combustion procedure that measures total nitrogen. When manure, compost, or fertilizer is measured, they can test for total nitrogen, nitrate, and/or ammonium. It is critical to know which test is being done.

Soil tests for potassium and phosphorus do not measure total nutrients. Instead, they use an extraction process that removes only some of the nutrient from the sample, and then this extraction is measured. The idea behind this is that the amount in the extraction is similar to what a plant can get out of soil, and so the reported results show the amount of K and P that is available to plants.

In the case of phosphorus, the extraction process removes soluble P and labile P (see phosphorus cycle) but not stable P. The actual amount measured depends very much on the extraction solution used, and these have common names such as Bray, Mehlich 3, and Olsen (bicarbonate). One problem with these tests is that each one gives different results. An optimum P level will give a 40 to 150 ppm for Mehlich 3, while for Olsen, it is 20 to 60 ppm. Be careful comparing results from different test methods and different labs.

This can be a real problem in books, blogs, and social media, where people report nutrient values without specifying the extraction procedure. Don't listen to someone who says a P value of 100 is too high. It might be high if measured using the Olsen method, but it is in the normal range if measured by Mehlich 3.

When manure, compost, or fertilizer is tested for K and P, the lab measures total nutrients. It is important to understand this because the nutrients reported for these products include those that are not yet mineralized and therefore not immediately available to plants.

CEC (cation exchange capacity) can be measured very accurately, but such tests are expensive. Most labs estimate the CEC by adding up the cations it measures. This approximation is quite good in acidic soil but not for alkaline soil, where the value is usually reported higher than it really is.

Using Soil Test Results

When ordering a soil test, you will be asked to specify the type of plant you're growing. The lab will then test your soil and make recommendations specific for that plant. This makes a lot of sense from an agricultural point of view. For example, if you are going to grow beans, the lab will make fertilizer recommendations to maximize the production of beans.

A market garden or a landscape situation is different. In these cases, you want to grow a wide variety of plants, and a lab can't provide a specific recommendation for each. At best, you now get a very general recommendation.

Most labs will provide you with a numerical result as well as a range chart showing values corresponding to very low, low, medium, high, and very high. The medium level is a good level for most plants. An extreme reading indicates either a deficiency or a toxic level.

Consider the example where phosphorus is very high, potassium is low, and all other nutrients are medium. The lab suggestion would be to add more potassium in the form of a synthetic fertilizer such as potassium nitrate. But you want to go organic. Now you need to find an organic source for potassium that does not contain phosphorus, and that is difficult. Wood ash is one, but it's becoming hard to come by, and it is alkaline. Composts and manures contain too much phosphorus.

This illustrates one of the limitations of a soil test. You might not be able to find an organic product that meets the recommendations, but at least you know that you should do your best to keep phosphorus out of the garden.

Maximize Yields

A tremendous amount of research has gone into soil testing and fertilizer recommendations. This is now a highly tuned process that needs to be understood. The goal of soil labs is to tell farmers how much fertilizer they should use to maximize profit, and there is nothing wrong with that. In a given field, a crop will produce a certain yield if no fertilizer is added. Add a bit, and the yield goes up. Add even more, and the yield will be even higher, but at some point, the incremental increase in yield reaches a plateau, where the cost of the fertilizer is more than the expected profit from increased yield. Soil test results try to find that sweet spot where the cost of fertilizer still produces profit.

That process works great for agriculture and for growing a specific crop. The system breaks down when you are trying to cultivate several different crops. You can't fertilize a field efficiently when every one needs a different fertilizer. At best, you use an average, which may or may not produce a profit on the total field.

The other issue is with landscapes that normally grow many different types of plants and where yield is not a fundamental goal. Adding extra fertilizer to make a spruce grow ½" taller is of no value. Soil testing labs do not have suitable recommendations for these situations and usually over-prescribe fertilizer. Most ornamental plants grow just fine in the low to medium range, but labs recommend medium to high ranges because of their focus on yield. In reality, there has been almost no testing to determine suitable ranges for ornamental gardens. We do not have good minimum levels that focus on healthy plants instead of yield.

Plants as Indicators of Soil Problems

A deficiency of a particular nutrient in soil will cause specific plant symptoms. This has led people to conclude that you can identify a deficiency by looking at your plants, but that does not work since the same symptom can be caused by different deficiencies. For example, symptoms for phosphorus deficiency may actually be a nitrogen deficiency. None of the memes and charts on social media can be used to confirm a deficiency.

Sweetgum with interveinal chlorosis.

Several other factors, including diseases, environmental conditions, insects, and pesticides, can produce symptoms that look like nutrient deficiencies.

A lack of iron can cause interveinal chlorosis, but so will all of these issues:

- manganese deficiency
- high soil pH
- zinc deficiency
- herbicide damage
- wet soil conditions
- compacted soil
- trunk-girdling roots
- plant competition
- high organic content in soil
- high salts
- high levels of phosphorus, copper, zinc, or manganese

Plant symptoms are a tool for predicting possible nutrient deficiencies, but in most cases, it is not a reliable one. Soil and plants are just too complicated, and problems can't be identified by a simple list of symptoms. Only a soil test or tissue test can identify a nutrient deficiency accurately.

Plant Tissue Analysis

Soil tests measure the amount of nutrients in the soil, but they tell you nothing about a plant's ability to use those nutrients. That is where plant tissue analysis can be useful. It takes a more holistic approach and looks for issues such as nutrient imbalances, toxicities, effect of pH on nutrient uptake, water availability, compaction, microbial activity, and other environmental conditions. It does not identify any particular issue, but it does signal a problem with the plants ability to take in nutrients.

Nutrient levels in plants vary depending on their type and stage of maturity. Samples analyzed early in the season allow time for making corrections. Late-season samples provide a picture into the past year's growth process and may allow you to evaluate any mid-season actions that were taken. It is also a good way to compare different sections of a field.

Taking plant samples is more complex than taking soil samples, and it is best to contact the lab and get specific instructions. Depending on the tests and type of plant, you will need to collect specific tissues at the right growth stage. You will also need to follow procedures to ensure that samples are not contaminated.

DIY Test Kits

There is a wide range of test kits and meters on the market. In general, the lower-priced items are of limited value. Higher-end products can be useful and provide the added benefit of allowing you to test soil more frequently than you would do using a lab.

Soil pH can be measured with three different ways: electronic meters, indicator test strips, and chemical colored dyes. In each case, you take some of your soil and mix it with water or a buffer solution. The soil mixture is then tested.

Since pH is a logarithmic scale, measures should include at least one decimal point. Most DIY kits do not meet this specification.

Chemical Colored Dyes

Colored dyes are combined with the soil-water mixture, and the resulting shade is compared to a supplied color chart to determine the pH level. These tests usually measure only whole numbers, and with some kits even that is a guess. These kits are almost useless.

pH Test Strips

Test strips for pH are advanced versions of litmus paper, and many people incorrectly still call them this. True litmus paper is extremely inaccurate and completely useless for measuring the pH level of soil. Test strips are more accurate since they have several color spots on each strip. To get reasonable results with these, there must be at least three color spots and a pH range of 5 to 10. Even with these specs, it is still difficult to measure much more accurately than one pH unit. These are low cost and very easy to use, so they are good for a quick field test.

Electronic pH Meter

A variety of garden pH meters are available. The probe that comes with them is inserted into the soil mixture, and the pH can be read directly from a display. The really cheap models come with a metallic probe that the instructions suggest you insert directly into soil. This is more convenient, but you will never get a useful reading without first making the soil mixture as described above. Even then the cheaper units provide only a very approximate reading. Better-quality meters will be supplied with a glass or plastic probe that measures pH more accurately. The lab version of these will measure to 0.1 pH accuracy.

Limitations of DIY pH Tests

The accuracy of the above options is limited, and that is a problem when you go to adjust your pH. Without an accurate value, you are just guessing as to how much amendment to add. The best option is to use a commercial lab or purchase a better-quality lab-grade pH meter with a proper glass probe. This will allow you to calibrate using standard puffers and get readings that are accurate enough.

Testing for Nutrients

Garden centers sell kits for testing nitrogen, phosphorus, and potassium. These kits are of limited value because they are not very accurate, and they do not give you an actual value for each nutrient. Instead you get a range value such as low, medium, and high. Without numerical values, it is impossible to calculate the right amount of fertilizer needed to reach a target value.

Some agricultural-grade colorimeters and better-quality reagents are available that will provide a more accurate value and can be useful in the field.

Except for nitrogen, nutrient levels don't change quickly, and for most purposes, an annual measurement is adequate. Using a commercial lab for testing may be a better option.

Determining Soil Texture

There are three options for determining soil texture: field method, jar method, and lab test. Anyone can easily do the first two and get reasonably values. The lab test is more accurate, although the accuracy for soil texture is not that critical since you won't be trying to make major changes to it. Knowing an approximate value is useful and will help you to understand potential soil problems.

In each of the following methods, remove any mulch, rocks, and large roots before you take a soil sample.

Field Method

Start with a feel test. Rub some moist soil between your fingers. Sand feels gritty; silt feels smooth; and clay is sticky. This will give you a very crude value.

Now try the ball squeeze test. Take some moistened soil and try to form a ball. Course soil such as sand and sandy loam breaks apart easily, and you won't be able to form a ball. Fine textured soil like clay and clay loam resists breaking apart, making it easy to form into a ball. Silty loams have characteristics half-way between these extremes.

The ribbon test is even more accurate, but takes a bit of practice to get the feel of it. Squeeze a moistened ball of soil between the thumb and fingers of one hand. See if you can make a single long ribbon out of the soil. Sandy soil won't form a ribbon; it just breaks apart. If you can form a ribbon that is at least 2" long, the soil has 20% or more clay. Shorter ribbons indicate more silt or small amounts of clay. Reference video: gardenfundamentals.com/soil-tests/.

Jar Method

Take a sample of soil and place it in a flat-bottomed jar that has clear sides. Fill the jar ¾ full with water and add one teaspoon (5 ml) of borax or a pinch of table salt. Give the jar a good shake, and set it down. The sand settles during the first minute. At that point, mark the height of this level. Wait 1 hour and make another mark to show the top of the silt.

Wait 24 hours and make a final mark to show the settled clay. Very fine clay may take longer to settle; waiting 48 hours may be required. Use a ruler to measure the height of each mark, and convert them to percent values. Now find the values on the Soil Texture Triangle (see page 16) to determine your soil's texture. Reference video: gardenfundamentals.com/soil-tests/.

Crusted Soil

A crusted soil exists when the surface becomes harder than the soil immediately below the top inch. It can be a result of chemical, biological, or physical conditions.

In normal conditions, rain and irrigation land on soil and slowly seep to its lower levels. Moving down, water carries salt ions with it, taking them out of the root zone. A chemical crust develops when the opposite happens. In hot, dry conditions, the rate of evaporation at the surface is higher than the rate at which water seeps deeper. This reverses the movement of salt ions, and they travel from lower levels to the surface of the soil. As water evaporates, it leaves behind a salt crust not unlike the crust in your tea kettle. These chemical crusts can be an issue with seed germination due to high salt concentrations.

Biological crusts are caused by living organisms growing on the surface, such as lichen, cyanobacteria, algae, and moss, which tend to exist in arid and semi-arid regions. Algae can also cause crusting in areas where water stands for several days at a time. Biological crusts play key ecological roles, including carbon and nitrogen fixation, and they help prevent soil erosion. They also add nutrients to the soil and may prevent seed from germinating. Depending on how they form, they might increase or decrease water infiltration.

The most common form of crusting is physical crusting, the result of rain or irrigation hitting the top layer of soil and disintegrat-

Soil Test: Soil Crusting

To check for crusting, have a look at the surface of the soil. Is the surface irregular, or does it look like a flat sheet? The flatness indicates possible crusting. Gently move some of the surface around with your fingers or a trowel. The crust will break up into sheets, and you will be able to pick up small sheets of soil. The hardness and thickness can be used to determine the severity of the problem. Thick, hard sheets indicate a high level of crusting.

Biological crusting will usually be green or green-blue. Chemical salt crust is usually white or tan.

ing the surface aggregation. This releases fine silt and clay particles that, when dry, pack down into a hard layer. The degree of crusting depends on other soil properties and on the droplet size of precipitation.

Soils that have a wide range of particle sizes, contain large amounts of fine particles, or have weak aggregation are more likely to develop crusts. High sodium levels result in weak aggregates and exasperate the problem. This crust reduces or prevents water infiltration, resulting in more runoff, which moves fine soil particles and accumulates them in low areas that are then more prone to crusting. Seedlings have trouble pushing through the hard crust, and gas exchange between the soil and air is reduced.

Eliminating physical crusting is fairly straightforward: prevent water from hitting the surface of the soil, develop strong aggregation by adding more organic matter, and reduce or eliminate any procedure that reduces aggregation.

Quantification of Microbes

Soil health is directly related to the number of microbes in your soil, and improving soil health is all about increasing these numbers. But there is a catch: How does the gardener count the number of microbes?

Some groups promote the idea of using a microscope to examine and count your microbes. They even sell courses and equipment for doing this. This can be a fun exercise, but it's not very useful for understanding your soil. Expert scientists have trouble doing this, and most microbes are still unidentified. What chance do you, as an amateur, have of doing this?

A huge variety of organisms exists in soil, and this population is constantly changing, making it difficult to quantify. Some labs do provide soil microbe testing, but for gardeners this will not provide much useful information.

There is no easy direct way for gardeners to quantify microbes, but some indirect ways can do it. Earthworms, which exist in many soils, are a good indicator of the microbe population since that is their main diet. Increase the number of microbes, and you will almost certainly increase the earthworm population.

The Tighty Whitie soil test is also a good and fun way to measure microbes. Microbes eat plant-based material, including cotton fibers. The speed at which microbes can consume cotton material is an approximate measure of the microbe population.

As microbes live and grow, they use up oxygen and give off carbon dioxide, just like humans. The amount of CO_2 produced is a direct

Soil Test: Tighty Whitie Soil Test

Take some white cotton underwear, weight it, and bury it in the garden. Your frilly silk panties may work in the bedroom, but not for this test. And don't use thongs—they just don't provide enough organic matter to give an accurate weight.

After five weeks, remove the underwear, and weigh it. How much of the cotton was consumed by the microbes? If most of it is still there, your soil is not very healthy, and you need to add compost. If you are left with the elastic, congratulations, your soil is in peak condition. Source: gardenfundamentals.com/tighty-whitie -soil-test-review.

> ### Soil Test: Soil Respiration
>
> Collect a fresh soil sample as directed by the instructions of the Solvita field respiration test kit. Insert the gel probe, and seal the supplied jar. Incubate at 68° to 75°F and out of direct sunlight for 24 hrs, and compare the gel probe with the supplied color index. For more information, see.nrcs.usda.gov/Internet/FSE_DOCU MENTS/nrcs142p2_053267.pdf.

measure of metabolic activity. Soil respiration can be measured by a qualified lab or a DIY Solvita test kit.

Aggregation also depends very heavily on the presence of microbes. They are essential for creating aggregates, and if their population decreases, aggregates start to disintegrate. The degree of aggregation is an indirect measurement of microbial activity. However, aggregation is a slow process, so it won't give you much information about short-term changes.

The level of organic matter is an indirect measurement of microbe activity. If you have fresh organic matter in the soil, microbes will find it. The more you have, the more microbes you will have.

Level of Organic Matter

The amount of organic matter in soil can be measured by a lab using one of two tests: the Walkley-Black method and weight loss on ignition method. The former is more accurate on soils with less than 2% organic matter, but it uses harsh chemicals, so many labs have stopped using it. The weight loss on ignition method is more eco-friendly, more popular, and better suited for soil with more than 6% organic matter.

These two tests produce different results. A soil with a 3% Walkley-Black level will get a 4.5% reading with weight on drying. Labs may use a correction factor to correlate the two methods. If you use more than one lab, make sure you know which method is used and whether or not they are using a correlation factor.

The organic matter level can also be estimated by looking at other soil properties. A high CEC indicates a high level of organic matter. Soils with a high bulk density tend to have low organic matter.

Compaction

Soil compaction occurs when soil particles are pressed together, reducing their pore space and increasing soil density. This reduces the space available for both air and water. Fine soil is easier to compact than course soil, and is even more susceptible to compaction when wet.

The soil property called *bulk density* is a measure of the degree of compaction, which increases as the volume of pore spaces decreases. In the field, compaction can be seen as puddling of water that has no place to go since the pores spaces are too small. Compacted soil becomes harder to dig, and when the soil is turned over, it tends to form clots instead of falling apart. Walking on the soil or riding over it with equipment will not leave tracks. Seeds have difficulty growing, and even larger plants are stunted due to a poor root system. In extreme cases, the lack of oxygen leads to anaerobic conditions along with the production of gases that smell like rotten eggs.

Soil compaction can be measured directly in the field using a penetrometer. As the device is pushed into soil, it measures the force needed to penetrate, mimicking a growing root. It is long enough to measure compaction both at the surface and at lower levels down to about 15" (40 cm).

Soil Test: Compaction

Homeowners can use a simple wire to get a relative measurement of compaction. Push a straight metal wire from a coat hanger or irrigation flag into the soil until it bends. A depth of one foot or more indicates good soil for root growth. Rocky soil will give you a false sense of compaction since the rocks stop the wire. Reference video: gardenfundamentals.com/soil-tests/.

Roots can penetrate soil with a pressure of up to about 300 psi. Above this, roots can no longer travel down, and they start to go sideways. Looking at roots of a mature plant can indicate the presence of a shallow hardpan.

Lab tests for bulk density and pore size distribution can be used to measure the degree of compaction. Since soil has varying densities, it is best to compare the area of concern to one close by that has been left in a more natural state.

Compaction is the result of applying pressure to soil. This squeezes aggregates together and collapses macropores, all the while pushing air out of the soil. It has the following effects on soil:

- Reduced air exchange, causing oxygen levels to drop and carbon dioxide levels to rise
- Reduced permeability
- More runoff, which leads to more erosion
- Less water available to roots
- Restricted root growth

Compaction is increased with regular cultivation and the use of heavy equipment. Annual plowing tends to decrease compaction, but can lead to the formation of a hardpan.

Hardpan

A pan or hardpan, defined as any layer of hard soil that is located below the surface of the soil, is a problem because it restricts water flow, air movement, and root growth. A *plowpan* is a pan made by a plow. A *claypan* or *fragipan* is a natural formation that consists of a hard layer of clay in the subsoil. A *duripan* is a hard layer caused by the accumulation of chemicals.

Dealing with natural hardpans is easier if you understand why they exist. This may require some specialized testing or contacting local agricultural agencies to better understand local conditions.

Plowpans can be reduced by plowing less and varying the depth of the plow. Deep plowing can be used to break them up, but they

Soil Test: Hardpan

The same test that is used for measuring compaction will also find a hardpan. In fact, with the wire test, it is difficult to determine if it is compaction or a hardpan.

You can also slowly dig a hole with a shovel or trowel. When and if you reach a hard layer, you have detected a hardpan, or you have reached an area with lots of stones. My soil has 8 to 10 inches of soil over top of a very gravely soil that becomes difficult to dig. A sandy, gravelly area may be difficult to dig, but it is not a hardpan and usually drains fairly well.

may return on their own if additional steps are not taken. A broad-fork is a manual tool designed for breaking up hardpans, but its penetration is not very deep. Incorporation of organic soil amendments can also be helpful. In certain cases, gypsum may help loosen up the fine clay particles in the hardpan. It is best to get some expert advice before doing this.

Drainage

Drainage is a general term that refers to the speed at which water leaves the surface or subsurface area of soil. Many gardeners use the term to refer to the rate at which water moves through the soil, but there are actually three different processes at play: *runoff*, *infiltration*, and *percolation*.

The best way to identify drainage issues is to watch the landscape during and after a heavy rain. It will allow you to see runoff as it's raining; any standing water after 24 hours indicates a drainage issue.

Runoff

Runoff is a process in which accumulating water runs along the ground and leaves the area. This happens most often on soil that is not flat, and such water tends to accumulate in low areas where it can't run any farther. Although this has more to do with the soil topography than characteristics of the soil, crusting and compaction increase runoff.

Excess runoff can have some beneficial effects. In areas with high rainfall, it might be advantageous to have most of it run away from the garden. The one problem with this is that water moving quickly carries with it surface soil, increasing erosion. It also tends to move the fine particles that accumulate at low spots, causing crusting.

In dry locations, runoff is a waste of a valuable resource, and every effort should be taken to keep rain in the garden. In fact, you might consider changing the topography to direct water to the garden.

Runoff can be reduced or eliminated in a number of ways:
• Change the surface topography.
• Mulch to slow the movement of water.
• Keep soil covered with plants.
• Use swales to direct water to drier areas.
• Slope terraces.
• Create stone barriers that run across the flow to slow it down.
• Increase infiltration and percolation.

Infiltration

Infiltration, the process of water moving from the surface into the soil, is affected most by surface characteristics. Crusting dramatically reduces infiltration because it creates a barrier at the surface of the soil that keeps water out. The large pores of sandy soil increase the infiltration rate, and clay soils decrease it. Aggregation creates very large pores and also increases infiltration.

Compaction decreases the rate of infiltration, so any steps to prevent or improve compaction will also improve infiltration.

Percolation

Percolation is similar to infiltration, but it has more to do with the water's movement once it enters the soil, and it measures how

Soil Test: Percolation Test

Dig a hole that is 1 foot (30 cm) wide and 1 foot deep. Fill with water and let it sit overnight so that the soil becomes saturated. Refill the hole.

Use a tape measure or ruler to measure the depth of the water. Remeasure the level every hour until the hole is empty. You may need to measure more frequently in quick-draining, sandy soil. Calculate an average hourly drainage rate.

A drainage rate of 1 to 3 inches per hour (2.5 to 7.5 cm) is acceptable. A value of less than 1 inch indicates a percolation that is too slow, and a value greater than 3 inches indicates a soil that drains too fast.

If the water does not leave the hole, or if it rises, it may indicate a high water table.

quickly it moves through soil levels. High percolation rates indicate that water moves quickly, usually the result of large pore sizes.

Percolation can be improved by increasing aggregation and decreasing compaction, which in turn produces larger pore spaces. The key is to increase the amount of air in soil so that it can be replaced with water.

A hardpan will also reduce percolation, especially if it is a shallow one.

CHAPTER 10

Gardening Techniques That Affect Soil

Most gardeners learn by following advice on how to do things. That is a great way to learn, but unfortunately, the description of the techniques almost never discusses their impact on soil. Rototill the soil to get rid of weeds; mulch to keep moisture in the soil; rotate your vegetable crops; and use companion planting are just some techniques. They all have a purpose, but how do they affect soil health? Understanding this will allow you decide if you should keep doing the procedure or if you should modify the way you do it. Don't do things in the garden because someone told you it is a good idea; do them because you know how they impact plant growth.

Tilling

Tilling the soil in spring, a time-honored tradition, can be done in ways that include rototilling, digging, cultivation, and using a broadfork. The perceived benefits include loosening up the soil, eliminating weeds, and incorporating soil amendments. It is interesting that agriculture has been moving away from tilling and adopting no-till practices, but most home gardens and small market gardens still rely on tilling. Many are not even aware of the move to no-till.

Tilling does loosen up the soil, temporarily. But this also breaks up soil aggregates, which in the long run, leads to issues such as crusting and compaction. Disturbing the soil also affects microbes directly. Larger animals like earthworms are killed, and the fungal hyphae are chopped up. The science on this is clear: tilling is not good for soil.

Plowing is a gentler way to loosen soil, and if done right, it can actually increase the presence of large aggregates in some soil.

Cultivation of soil adds more air into the top layer of soil, thereby increasing microbial activity, which results in a more rapid depletion of OM, one of the main reasons why traditional agricultural fields have low organic levels. Cultivation speeds up its decomposition to a point where the loss is greater than the addition of new organic matter. In the long term, this results in a loss of aggregation and microbial activity.

The difference in available nitrogen in a cultivated field and a no-till field can be dramatic. Due to higher reserves of organic matter, the no-till field can supply up to 5 times as much nitrogen. Tilling also changes the form the available nitrogen takes. In the presence of higher oxygen levels, the ammonium cation is more easily oxidized to nitrate, an anion. Because nitrate has a negative charge, it does not stick to soil particles or organic matter as well as ammonium and is much more likely to be washed away as water percolates through soil.

The reduction of weeds from tilling is also a short-term event. Soil lower down contains vast amounts of seeds, known as the soil seed bank. They remain dormant because of a lack of light, but cultivation brings them to the surface to germinate. It is true that cultivation decreases some of the growing weeds, but it also makes ideal conditions for new weeds to grow from the seed bank. Tilling can also increase the number of perennial weeds by cutting rhizomes into smaller pieces which then start to grow.

Historically, double digging has also been popular, especially for starting a new garden. The top layer of soil is removed along a trench. The subsoil is then loosened, turned, and organic matter added. The topsoil from the next trench is then turned onto the first trench and so on. This results in very loose soil that is highly mixed. The problem with this method is that it suffers from all of the above-mentioned issues. It might be suitable for soil that is in terrible condition, but it should not be used for most soil.

The bottom line is that you should disturb soil as little as possible.

Working the Land

Any weight that is placed on soil will cause compaction. This includes walking or driving equipment over it; all such activities should be kept to a minimum. Wet soil and fine soil, like clay, are more prone to compaction.

One of the benefits of raised beds is that they are built so that foot traffic stays in the paths, eliminating compaction in the growing areas.

The following will reduce compaction caused by equipment:

• Reduce tire pressure.
• Use larger tires to spread out the weight.
• Use the same lanes each season.
• Maintain proper tire pressure.
• Use the lightest equipment that can do the job.
• Combine operations to reduce the number of trips.

Mulching

Mulch, any material that covers the ground between plants, will retain moisture, suppress weeds, and keep the soil cool; some people feel it makes the garden look better. It can be either organic or inorganic, but organic material has the added bonus of improving the soil as it decomposes. The advantage of inorganic material is that it does not need to be replaced.

Benefits of Mulch

Mulch reduces the evaporation rate at the surface, which helps maintain moisture throughout the soil, which in turn reduces the need to irrigate. Plants have a constant supply of water at their roots, and microbe populations have less interruption.

Plant roots grow best in cool conditions; when it gets too warm, they stop growing new roots and reduce the absorption of nutrients. A mulch layer keeps the surface of the soil cooler, allowing roots to grow throughout the season.

Most annual weeds are produced when their seeds from the soil bank are exposed to light near the surface of the soil. Mulch prevents light from reaching them and keeps them from germinating. New seeds from plants tend to be washed through the mulch to lower levels where they won't germinate.

There is also another process at play here. Many mulches have a high C:N ratio, which means that microbes need to take nitrogen from the surface of the soil to decompose the mulch. This reduces the nutrient levels right at the point where weed seeds are trying to germinate. Indirectly, microbes help starve the seedlings.

Plants grow better under mulch. One study that looked at tree growth under wood mulch found that the tree mass was 170% higher than similar trees grown without mulch. This is due to the benefits mentioned above as well as to extra nutrients provided by the decomposing organic mulch.

Effect of Mulch on Soil

As organic mulch decomposes, it has a significant effect on the soil. In a five-year study that compared the effect of fertilizer, compost, and wood chip mulch on soil by measuring density (i.e., compaction), moisture, organic matter, respiration (microbe activity), pH, nitrogen, phosphorus, and potassium, wood chip mulch increased the organic matter in soil, decreased compaction, increased microbial activity, and increased nutrient levels. These changes lead to better aggregation and better plant growth.

Mulch also reduces crusting, increases rain infiltration, and reduces runoff. Raindrops can no longer hit the surface of the soil, eliminating the crusting problem. The mulch also spreads out the rain, allowing it to more slowly percolate to the surface of the soil, where it is easily absorbed.

Effect of fertilizer and mulch on soil

	Fertilizer	Compost	Wood Chips
Density	Lower +	Lower ++	Lower ++
Moisture	Same	Up +	Up ++
Organic matter	Same	Up ++	Up +
Respiration	Up +	Up ++	Up ++
pH	Same	Up ++	Up +
Nitrogen	Same	Up ++	Up +
Phosphorus	Up +	Up +++	Up +
Potassium	Same	Up +++	Up ++

Source: GardenMyths.com
based on data by Bryant C. Scharenbroch and Gary W. Watson

Different mulches.

Compost

Compost is very good mulch that provides many nutrients and is very effective at improving the quality of soil. It is generally applied as a thin layer of about one inch and, as such, is not great for reducing weed growth. It is better to apply compost as mulch than to dig it into the soil.

Wood Chips

Also called arborist wood chips or wood mulch, wood chips are an excellent mulch for landscapes where there is relatively little disturbance to the soil. It can also be used for growing vegetables, but some extra care needs to be taken to ensure that the wood is not buried in the soil, where it can cause a temporary nitrogen deficiency. If using it in a vegetable garden in cold climates, rake it aside in late fall to allow the soil to warm up in spring, ready for planting. Once plants are a few inches tall, the mulch can be replaced.

Soil Myth: There Is No Such Thing as Too Much Compost

Compost is organic and slowly provides nutrients to the garden. Since it is good for the garden, many people feel that more is better, but that is not true.

The problem lies with the nutrient ratios in compost compared to the nutrients used by plants. Plants use NPK in a ratio of about 7-1-6. They use seven times as much nitrogen as phosphorus. Compost generally has an NPK ratio of 1-1-1, namely equal amounts of each nutrient. If you supply a plant with the correct amount of phosphorus using only compost, it will be lacking in nitrogen. If you use more so that the plant has enough nitrogen, you will be supplying way too much phosphorus, which is what many people do. After a few years, their soils become toxic due to the high levels of phosphorus.

Small amounts of compost are good for the garden and will avoid toxic P levels.

One benefit of wood chips is that they only need to be applied every three or four years.

Hay and Straw

These make excellent mulches in a vegetable garden. They can be applied thickly enough to prevent weeds, while still providing good air circulation to the soil, and they are great for preventing soil-borne diseases from being splashed up onto plants. If you have not heard about Ruth Stout, the Mulch Queen, search out some information on her. She was a big fan of straw mulch.

Hay is a fresh cut and tends to have a higher nutrient value but can also contain more seeds. Straw, the lower cut done after grain is harvested contains fewer weed seeds and has low nutritional value but tends to last longer. With a thick enough layer, seeds are not really an issue.

I find that straw lasts a couple of years before it needs to be renewed. It is the best choice for vegetable gardens.

Plant Refuse

Plant refuse includes things like fall leaves, pulled weeds, and old plant material. It can also be the by-product from a local industry such as cocoa husks or spent coffee grounds. All of these can be composted first, but they can also be used as mulch where nature will do the composting for you. The key here is to use something local, not something that needs to be trucked long distances.

Paper Products

Newspaper and cardboard have become popular items for mulching. They are a waste product that can be recycled in the garden. Although the intent sounds good, they are not a great choice because they prevent water and air from reaching plant roots, and they add very few nutrients.

Black Plastic

Black or clear plastic is popular in vegetable gardens. It certainly controls weeds, but it also heats up the soil, which is not good for plants later in the season. If you use it, make sure you install an irrigation system under the plastic so plants can be watered.

Growing in plastic has become a very popular agricultural practice. It has no long-term benefit for the soil, and it increases our waste and pollution problem. Although it can be a time-saver, it is not an ecologically good choice for smaller operations.

Landscape Fabric

Landscape fabric, also called weed barrier, is a black plastic that is manufactured with small holes in it that, it is claimed, air and water pass through. This is partially correct, but much of the water that lands on it will run off. It is normally covered with some type of mulch because it looks ugly.

It has no place in the landscape. It does stop weeds for a year or two, but very soon you find seedlings growing on top of it. Plant roots also grow through it, and when you try to remove it, you damage their roots. It has limited use in a vegetable garden. It does work better in a nursery area or on walkways between beds, where it keeps weeds from growing, provided it is not covered.

Stones

Stones can be an effective mulch. They keep the soil cool, reduce evaporation, and last a long time. Unless you use a thick layer, they do not keep weeds down, and they don't improve soil health. Don't use a landscape fabric under stones.

Hoeing

This is a long-time practice that people find hard to give up. Hoeing is reported to do two things: get rid of weeds and break up soil crust. Hoeing does get rid of crusting, but it soon returns. The reason is that crusting is a result of a loss of aggregation, and hoeing destroys aggregates. So hoeing actually makes crusting worse in the long run.

Growing weeds are removed, but the disturbance to the soil brings new weed seeds to the surface. This is why hoeing must be repeated all summer long. This heavy workload can be replaced with mulch.

Cover Crops

If you have a look in nature, you realize that most bare soil is quickly covered with plants, and this plays a critical role in maintaining healthy soil. When soil becomes bare, erosion increases, organic levels drop, and temperatures rise, resulting in less microbial activities in the top layers.

Traditional agriculture and gardening in temperate climates results in bare soil most of late fall, winter, and early spring, and is one reason our agricultural fields have become less productive. Cover crops, also called green manure crops, are used to alleviate these problems. The goal is to have something growing all year long. As soon as a crop is harvested, it is replaced with another plant. These cover crops provide many benefits for soil.

Reduced Soil Erosion

Cover crops prevent wind and rain from removing the top layer of soil by softening the effect of raindrops and reducing wind from reaching the surface of the soil. The roots also bind and stabilize soil particles.

Snow adds some protection in winter, but when it melts, the excess water washes nutrients out of the soil. Cover crops reduce run-off and absorb nutrients before they are washed away.

Weed Suppression

As soon as a crop is harvested, weeds start growing to fill in the void. A cover crop of fast-growing plants will both shade weed seedlings and compete with them for nutrients, reducing the weed pressure for a future crop. Buckwheat, a good choice for midsummer, provides flowers for bees. The key is to use lots of seed to create a very dense cover. Consider using as much as five times the normal seeding rate, and also mix in other types of seeds.

Increase in Organic Matter

In many cases, cover crops are not harvested and are grown simply to improve the soil. Before the next crop is planted, they are killed by either cold or manual techniques, and the crop residue is left on the soil for nature to take over.

The amount of organic matter added to the soil depends on the type of plant used, the length of time it was growing, and how it is harvested. In Southwestern Ontario (zone 5), oats and rye can contribute anywhere from 900 to 4,500 lbs/ac (1,000 to 5,000 kg/ha) of dry, above-ground residues. These numbers will be higher in warmer climates.

It is estimated that only 20% of plant residue is added to the soil organic matter. The remaining 80% becomes part of living organisms or is released as gases. At the above rate, if no other organic matter was added, it would take 20 years to add 1% of soil organic matter, which seems very slow; but consider the fact that without a cover crop, the value would actually be dropping.

Increase in Nitrogen Levels

Legumes are common cover crops, and if they are completely added to the soil, they inject fixed nitrogen for the next crop. If the seeds of the legume are harvested, the remaining plant residue adds very little nitrogen to soil since most of the nitrogen is in the seeds at time of harvest.

Nitrogen catch crops are used to scavenge any remaining nitrogen in fall and hold on to it over winter, when it can be released in time for next season's main crop. Clover and ryegrass work well for this.

Reduce Pests

A cover crop can be an alternate host for a pest and consequently reduce the pest before main crops are planted. Marigolds, planted immediately before root crops, will reduce root rot nematodes. The nematodes are attracted to the marigold roots, but are unable to reproduce in them, thereby reducing their numbers in soil. This works only if both crops are grown in the same season, one right after the other.

Cover crops can also provide a habitat for beneficial insects such as predatory mites and ladybugs.

Improve Soil Structure

Keeping roots growing in the soil all year long improves aggregation and soil structure. It also prevents aggregates on the surface of the soil from being damaged by irrigation and rain, resulting in less crusting.

Cover crops like daikon radishes, which produce large roots, can be very effective at penetrating hard soils. After they die, they also leave loads of organic matter right in the soil.

Water Management

The above-ground portion of plants reduces the force of rain and helps it soak into the ground, reducing runoff. The newly killed plant residue contains a significant amount of moisture that is then incorporated into the soil.

In wet areas, a growing cover crop can speed up the drying of soil in spring and provide an earlier planting time for the main crop.

Raised Bed Gardening

Traditional farming uses flat fields where planting can be done anywhere. As crops are rotated, different spacing can be used between rows. This provides a very flexible system that works well when the field is managed with tractors.

As you move to smaller fields and replace the automated equipment with manual or semi-automated equipment, raised beds become an attractive option. They can take two forms. A simple raised bed is one where the soil from the walking pathways has been moved onto the growing area. These beds tend to be about six inches higher than the pathways. Another form is to enclose the raised bed with walls, usually made from wood, or concrete blocks, but other material can also be used.

Raised beds work well for smaller gardens and market gardens up to about an acre. They have the benefit that the location of the beds is permanent and is never walked on. This significantly reduces compaction in the growing area.

It is also more efficient to amend a raised bed since all of the additives go into the growing areas, and none are wasted on the paths. You can also have different types of soil or amendments in different beds, depending on the plants that will be grown there.

Proponents of walled raised beds make numerous claims for them, but many of these claims also apply to raised beds without walls, or even to growing on flat ground. Walled beds look neater, and if high enough, they can be more comfortable to work in. They can also be used in places with very poor natural soil and even on areas without any soil below them. The walls may reduce some pest issues like slugs and snails and may keep out small rodents.

Benefits of a Raised Beds (no walls)
- Provide better drainage, provided their soil is similar to the ground under them. If the soil in beds is very different, they can create a perched water table, making the bed drain poorly.
- Dry out and warm up quicker in spring, allowing earlier planting. On the flip side, they cool down more quickly in fall.
- Less compaction from foot traffic.

One of the main benefits of raised beds is that they have changed the pattern for planting. Instead of using the traditional wide space between rows found in agriculture, they are planted with narrow rows or even with no rows at all. In effect, you replace many walking paths with a few and move the rows closer together. This works only because the planting areas have little or no compaction and are more heavily amended. You need healthy soil to plant closer together.

Crop Rotation

Crop rotation is the practice of growing a different type of plant each year in a particular location. There are many schemes for this that use between a 2-year to a 10-year rotation and may include a year when only a cover crop is grown. Historically, a field was left fallow every few years, but it is much better to use it for a cover crop.

The way in which plant type is defined can vary. Some do it by plant families so that all the brassicas are one type. Corn and legumes are two other common families. Others separate root crops from non-root crops so that they never have two root crops in subsequent years.

Many studies have shown that yields improve with crop rotation (10% to 25%), but the reasons for this are still not clear. It is commonly stated that crop rotation prevents one crop from depleting soil of a particular nutrient. Since all plants use the same nutrients in more or less the same amounts, it is unlikely that such a simple explanation is correct. Besides, research shows there is still an improvement even when the nutrient levels are kept high with fertilizer.

A common suggestion is to plant legumes right after corn. Corn is a heavy feeder of nitrogen, and the legume adds nitrogen back to the soil. The problem with this idea is that at harvest time most of the nitrogen is in the legume seed, so that if the crop is harvested, very little nitrogen is added to soil. Even when lots of nitrogen fertilizer is used, corn following corn produces a lower yield than corn following legumes. The benefits of rotation can't be explained simply on the basis of nutrient availability.

It is possible that the previous crop changes soil pH. For example, growing fava beans may acidify soil enough to free up more phosphorus, which the following corn crop needs in abundance. The previous planting could also change the biodiversity of the microbes that provides benefits for a future crop. Plants have different root structures that penetrate soil differently. Rotated crops are more drought resistant and use nitrogen more efficiently. Soil health, measured in a variety of ways, improves with crop rotation, but scientists do not understand why all of this happens.

Crop rotation also reduces the pressure of pests and diseases. Growing the same crop in the same soil repeatedly will make it easier for pests and diseases to locate near the crop and be ready to infect in future years. Crop rotation interrupts this cycle.

The length of the rotation cycle is important for pest control. A two-year cycle works well to reduce leaf blight on onions, but a seven-year cycle is needed to reduce clubroot in radish and turnip. It is also essential that non-host crops are used during the off years.

Weeds grow better beside certain crops. Soybeans planted after wheat have fewer weeds because of the allelopathic effects of the wheat.

Crop rotation makes sense for larger areas but has limited value in normal backyards. The plants are too close together to have much of an impact on pest control, and most of these gardens are more heavily amended with organic material, so many of the benefits of crop rotation are lost.

Companion Planting

Companion planting is a very popular practice in which two or more plants are grown together so that at least one benefits from the presence of the other. The most common example of this is the Three Sisters agriculture system of growing corn, beans, and squash together. It is claimed that the beans fix nitrogen that helps the corn to grow; the corn provides support for the bean; and the squash shades the ground to keep weeds down and moisture in the ground.

It is true that North American natives did use this gardening technique, but they did it for convenience, not for high production rates. The reality is that beans keep almost all of the fixed nitrogen for themselves and don't make it available to the corn. The corn shades and competes with the beans. The squash does shade the soil, but uses a lot of the moisture for itself. Using today's cultivars, the Three Sisters system is not more productive than monocultures.

Most of the common companion planting schemes presented in books and online have no scientific basis. They have been invented by writers and repeated by the general public. There are some well-researched combinations that do work, but they are all agriculture based. Before spending time on trying companion planting, research the specific combination that interests you, and try to find some documented evidence that it does work.

Using companion plants to reduce pests is a common practice. Unfortunately, most research in this area focuses on the number of pests, when it should be focused on overall yield and quality of yield. A companion plant may reduce the number of a particular pest, but if the yield does not go up, or the quality doesn't improve, it has no real benefit.

CHAPTER 11

Solving Chemical Issues

Soil chemistry is critical to plant growth, and yet it is poorly understood by most gardeners. They have been conditioned to think that plants need to be fertilized in order to grow. They are in constant search for that right combination of chemicals to grow better plants, without ever taking the time to really understand what goes on in the soil on a molecular level. As a result, many practices do little for the health of the plant, soil, or environment.

This subject is also rife with myths and misunderstandings. Take the time to really understand what you are doing before making any chemical additions to soil.

Buffer Capacity

Changing soil pH is not a simple matter of just adding some acid to make it more acidic or adding lime to make it more alkaline. Many factors contribute to soil pH, and changing it is complex.

As an example, consider a soil where the pH is too acidic and you would like to make it more alkaline. The common recommendation is to add lime. This sounds simple. Take some lime, add it to soil, and do another soil test. What you find is that the pH has not changed. The reason is due to a soil property called *buffer capacity*, which allows it to absorb either alkaline or acidic material and maintain its pH.

What causes this buffering? In alkaline soils, oxides and carbonates of calcium, magnesium, and potassium neutralize any added acid and keep the soil pH from changing. In acidic soil, aluminum oxides and iron hydroxides do the same thing for added alkaline material, like lime. In neutral soil, OM and mineral weathering play a big role in maintaining pH.

Each soil has a different ability to absorb acids and bases: its buffer capacity. Soil with a high buffer capacity can absorb more acid or base without changing pH than soil with a low buffer capacity. High-buffer soils tend to have more clay, higher levels of organic matter, and higher CEC. A high buffer capacity is generally good for plant growth because it keeps the pH steady, but it can be a problem when you are trying to adjust soil pH.

Increasing pH

The ideal pH range for plants growing in mineral soils (those made from rock) is 6 to 7, but for organic soils (peat and marsh bogs) a better range is 5.5 to 6. The range also depends on the type of plants. Acid-loving plants like blueberries and rhododendrons want a pH around 5.5.

Peat soils and tropical soils are very difficult to change, but the pH can be increased in mineral soils by adding lime. Strictly speaking, lime is calcium oxide or calcium hydroxide, but the term is also used to describe a wide range of calcium-containing compounds. Agricultural lime is usually calcium carbonate, or limestone. All of these soil conditioners will neutralize acids and increase pH.

Before adding lime to soil, you must get a soil test done to determine the buffer pH. Knowing this value, the lab can then calculate the correct amount of lime to add for the pH change required.

Knowing the calcium and magnesium levels in your soil will also help you select the correct liming agent. For example, dolomitic lime should not be used on soil that is high in magnesium since it raises the magnesium level. In this case, gypsum would be a better option since it contains calcium sulfate and no magnesium.

Soil Myth: Add Lime Based on pH

When soil is too acidic, the common advice is to add lime, and the recommended amounts are usually based on the pH of the existing soil. Specific amounts are given for each pH unit that needs to be increased. For example, a well-known university recommends "four tablespoons of lime per square foot, to raise the pH by two points." This advice is almost always wrong.

The current pH value is used to determine if a change in pH is required, but the value is not used to determine how much lime is needed. Instead, the buffer pH value determines how much lime is required, and this value can be determined only by a proper lab test. The correct amount of liming agent depends on the product that will be used since each one has a different capacity for changing pH.

Since simple pH test kits don't provide a value for buffer capacity, they should not be used to change pH.

Different liming products also change the pH at different rates. Dolomitic lime is slow-acting, needing a couple of years to take full effect. The benefit of slow-acting is that it also lasts longer before it needs to be reapplied.

Decreasing pH

Decreasing the pH of alkaline soil can be very difficult to do, and in many cases it is not necessary. Although plants prefer a pH below 7, most will grow just fine at a pH of 8. Keep in mind that measured pH values are based on the total soil profile, and the pH in the rhizosphere can be different by as much as 2 pH units. The soil might be alkaline, but the plants condition the rhizosphere to be acidic.

Acid-loving plants like rhododendrons, azaleas, and blueberries will just not do well in a pH above 7, but others actually prefer a more alkaline condition. Except for acid-lovers, almost everything grows well in my 7.4 soil.

Calcareous soil is difficult to change because it normally contains minerals, like limestone, that are constantly making the soil alkaline. As acid is added to the soil, it is quickly neutralized as more limestone solubilizes. Even acidic rain at a pH of 5.5 has no effect. Peat moss has been recommended for acidifying small areas, but the pH drop will last for only about a week.

Fertilizer can be used to acidify soil; ammonium sulfate and sulfur-coated urea are good options for this. For lowering pH even further use agricultural sulfur. It is relatively inexpensive and slow-acting, but it's effective for only a short period of time. Bacteria convert sulfur to sulphuric acid. Powder is harder to handle but works faster than granular material. It needs to be reapplied annually, and the amount required depends on several factors including soil texture. It is best to follow soil test recommendations.

Aluminum or iron sulphate should not be used to lower pH since the aluminum and iron can become toxic, and they are less effective on a pound-by-pound basis.

Saline and Sodic Soils

A saline soil has a high level of soluble ions, including sodium, calcium, and magnesium. In normal soil, the concentration of ions is lower in the soil solution and higher inside the roots. In saline soil, the opposite is true. This makes it very difficult for water to flow into roots, and in severe cases, water will flow out of the root and into the soil. A plant in such conditions can be surrounded with lots of water and die from dehydration.

This condition is usually a problem in dry, warm regions where evaporation at the surface of the soil causes water and salts to move up from lower levels. The salts are then deposited at or near the surface, next to plant roots, and you may see a white crust. Saline soil has a conductivity above 4 mmhos/cm and a pH below 8.5. Sensitive plants are harmed at anything above 2 mmhos/cm.

Since these salts are soluble, it is best to water more and reduce evaporation rates, which will wash the salts down below the root zone. Drip irrigation and mulching both help.

Sodic soil has high levels of sodium and low levels of calcium and magnesium. Excess sodium combines with the negative charges on clay, which disrupts its structure: clay falls apart into extremely small particles that pack close together and fill pore spaces. These soils become difficult to till and have poor seed germination, and plants have difficulty growing.

The cause of sodic soil can be similar to saline soil, but it can also result from a high seawater table and irrigation with seawater. Sodic soil has a high sodium level, a conductivity below 4 mmhos/cm, and a pH greater than 8.5.

Reclaiming sodic soil starts with the addition of gypsum to introduce significant amounts of calcium that will replace the sodium attached to clay particles. The soil can then be irrigated to remove the excess sodium. It is also critical to fix any high-water-table problems.

The following steps will help prevent or reduce saline and sodic conditions:

- Level out low spots so salts can't collect.
- Use high-quality irrigation water.
- Keep soil moist to reduce movement of salts to the surface.
- Irrigate more to leach out salts.
- Keep fertilizer to a minimum.
- Test soil more frequently to detect problems early.
- Plant crops on ridges or use furrow-irrigated fields.
- Use drip irrigation for more even watering.

Soil Myth: Gypsum Improves the Tilth of Clay Soil

Heavy clay soil is common, and it is very hard to dig in such soil. A popular suggestion for solving this problem is to add gypsum, which is reported to loosen up the clay, make it easier to work, and improve its structure. In fact, gypsum has little or no effect on clay soil unless it has a high sodium content.

Increasing CEC

Sandy soils have a low CEC (cation exchange capacity), and as a result, nutrients are quickly washed out of the soil by rain and irrigation. Higher CEC values indicate very nutritious soil because it has many charged sites that can hold on to nutrients until plants are ready for them. Increasing CEC increases the nutritive value of soil.

There are three ways to increase CEC:

- Increase the clay content
- Increase the organic matter level
- Increase the pH

Increasing clay content is not very practical since it requires large amounts of clay, but it may be an option.

Organic matter levels can be increased in various ways: mulching, cover crops, addition of compost, reduction in cultivation, to name a few. The greatest effect on CEC comes from older, highly decomposed humus material, making this a long-term process.

Highly acidic soil can be limed to increase the pH. This removes the hydrogen ions from charged sites on soil and OM, allowing them to be replaced with nutrient cations.

One popular idea is that managing the Ca:Mg or Ca:K ratios is significant for plant growth, but this is not supported by science. It is more beneficial to increase the level of these ions in nutrient-poor soils rather than worry about their ratio. As the amounts increase, so does the CEC.

Synthetic vs Organic Fertilizers

Much has been written about the differences between synthetic fertilizer and organic fertilizer, especially by pro-organic groups. Many of these ideas are simply wrong. The term *organic* here refers to natural products that are allowed in certified organic farming. Synthetic fertilizer is any human-made fertilizer—such as ammonium nitrate, calcium sulfate, and potassium phosphate—including the organic chemicals like urea, because they are human-made. These products easily dissolve in water and are almost immediately available to plants.

Organic fertilizers are more natural even if they have been manipulated by humans. Manure and compost are good examples, but even rock dust, which is mined and processed, is considered to be organic. Bone meal, blood meal, and soybean meal are also processed and considered organic. These fertilizers consist of both free nutrient ions and larger molecules. The free ions are readily available to plants, but they represent a small percentage of the total nutrients. The major component of most organic products are large molecules that need further decomposition before plants can use them.

A major difference between synthetic and organic fertilizer is that synthetic releases nutrients to plants quickly and organic is a slow, long-term feed. This quick-feed characteristic of synthetic fertilizers is seen as a bad thing in some circles, but it can be just want plants need. If you are growing crops in a short season and you want to get as much growth as you can, then synthetic fertilizer is a good choice. Ornamental gardens, on the other hand, have no rush to produce large plants, so organic fertilizers work better.

When a garden has plenty of nutrients but is lacking in one, it is best, from both a cost and an environmental perspective, to just add that one. This is easier to do with synthetic fertilizers since they can be selected to add the missing nutrient. Most organic fertilizer supplies a wide range of nutrients, and in many cases, like manure and compost, you don't even have a good handle on what is in them. How much magnesium does your compost have?

Synthetic fertilizer is also easier to handle. If you want to use manure (1-1-1) to supply nitrogen, you need the equivalent of twenty-six 50-poundbags to add the equivalent of one 50-pound bag of 26-5-10.

Another major difference between synthetic and organic fertilizer is that the former does little to increase soil aggregation or CEC levels, while the latter does. The more highly processed an organic fertilizer is, the less effective it is for soil development. Highly processed fish fertilizer or blood meal does little to improve soil, compared to manure, plant refuse, and compost.

Understanding Fertilizers

Each country has its own requirements for labeling fertilizer; most include values for N, P, and K. Some countries, like Australia, also report the sulfur (S) level. The letters N, P, and K are the elemental symbols chemists use as a shorthand to describe the chemicals nitrogen, phosphorus, and potassium. (K stands for *kalium*, the Latin name for potassium.) If you have trouble remembering whether P stands for phosphorus or potassium, remember that these nutrients are listed in alphabetical order. Phosphorus comes before potassium in the alphabet, and so P comes before K.

Soil Myth: NPK Is the Percentage of Nitrogen, Phosphorus, and Potassium

The N value is the % nitrogen.

The P and K values are the % P_2O_5 (phosphoric acid) and % K_2O (potash) and *not* the % P and % K as so many references claim.

The following will help you convert to % P and % K:

- P_2O_5 consists of 56.4% oxygen and 43.6% phosphorus by weight. To get the % P value, multiply the reported NPK value by 0.436, or approximately half of the reported value.
- K_2O consists of 17% oxygen and 83% potassium by weight. To get the % K value, multiply the reported NPK value by 0.83.

Using this information, you can see that a fertilizer NPK number of 10-10-10 contains 10% nitrogen, 4.36% phosphorus, and 8.3% potassium. These conversion numbers will help you determine the correct amount of fertilizer to add to your garden so that you meet the soil test recommendations.

Nutrient Availability

The nitrogen value in NPK measures the total amount of nitrogen available. For most synthetic products, the nitrogen is available as soon as it dissolves in water. In the case of urea, microbes have to first convert it to useable nitrogen, but this happens very quickly in soil. Urea also vaporizes fairly quickly, so urea fertilizers should be watered in, or cultivated below the surface of the soil, so the nitrogen is not lost to the air.

The nitrogen in organic sources is mostly in a slow-release form. As much as 97% of compost is tied up in large molecules and not immediately available to plants. Most of the phosphorus in compost is also tied up in large molecules and is only slowly released to soil. Once released, it quickly complexes with cations and gets absorbed to soil.

Potassium does not get incorporated into large molecules and is released from organic sources fairly quickly in the form of a potassium ion.

Effects on Soil pH

The type of fertilizer used affects soil pH. Those containing ammonium ions, such as ammonium phosphate, acidify soil. As organic matter breaks down, it releases nitrogen as ammonium ions and also acidifies soil, as does urea. Nitrate fertilizer tends to increase soil pH.

Don't Feed Plants

It is a common misconception that we fertilize plants. This belief has been promoted by many gardening and farming experts, and it is even supported by fertilizer companies. A very common question is, Which fertilizer should I use for plant X? Boxes of fertilizer for home gardens are even sold for specific plants.

If you want to have some fun, use the image feature of Google and look for tomato fertilizer. You will see dozens of such specialty fertilizers, all of which have different formulations. How can all of these products be the best tomato fertilizer? They can't. At most,

Soil Myth: Balanced Fertilizer

When commercial fertilizer first became available, manufacturers developed the idea of a "balanced fertilizer." This seemed to make a lot of sense. It was known that plants need N, P, and K, so why not give it to plants in equal amounts, i.e., a balanced fertilizer.

Products labeled with fertilizer numbers like 5-5-5 or 10-10-10 became very popular with home gardeners. You could not go wrong since you were providing all three of the main nutrients. As explained in another myth, this so-called balanced fertilizer is not even balanced, since a 10-10-10 is actually a 10-4-8.

We now know that plants don't use the nutrients in equal amounts, so the idea of a balanced fertilizer does not make any sense. Unfortunately, it is still frequently recommended.

only one can be the right product, but in reality, none are correct for your soil.

We don't fertilize plants. The purpose of fertilizing is to replace missing nutrients in soil. If your soil has adequate amounts of all nutrients except nitrogen, you need to add nitrogen. Someone across town might have lots of nitrogen but be low in potassium. They need to add potassium. This is true even if you are both growing tomatoes.

Fertilizing is all about adding the missing nutrients, not about feeding plants. The goal is to build up the reserves in the soil so all nutrients are available when plants need them.

The next question is, What nutrients are missing? Only a soil test can tell you that.

Synthetic Fertilizers

Industrialized societies have been conditioned to believe that plants will not grow without fertilizer, and that is simply not true. Agriculture adds fertilizer to increase yields, not to make crops grow. Most could grow crops without fertilizer, but adding it increases yield and therefore profit. The fertilizer recommendations they use are based on lab tests, but unfortunately, many small growers and most home-owners simply fertilize because they think they need it.

If a soil test indicates a deficiency, add just the nutrients that will fix the deficiency. If there is no deficiency, don't fertilize. If you have not done a soil test, assume there is no deficiency, and don't fertilize.

Most landscapes and flower gardens need no fertilizer. Vegetable gardens on established healthy soil do not need to be fertilized because they usually receive enough organic matter to provide low levels of nutrients.

Market gardens are half-way between home gardens and regular agriculture. Profit is important, and fertilizer may increase yields. Fertilizing should be done following a soil test.

In all cases, you should take into account the organic matter that is being added to the garden, both for this year and the previous five years. It can easily supply most if not all of the nutrients needed by plants.

Commercial Chelates

Chelated fertilizers have become more popular and are commonly used for an iron deficiency; however, selecting the right product is essential. Each chelate compound delivers a different amount of ion and is effective at different pH levels.

- Iron—EDTA: Releases the most iron to the soil, but it is good only in acidic soil. It is the common form found in most liquid fertilizers. EDTA also has a high affinity for calcium and should not be used on soil with a high calcium level since it becomes ineffective at protecting iron.
- Iron—DTPA: Effective up to a pH of 7.5 and not as sensitive to calcium as iron—EDTA.
- Iron—EDDHA: Can be used up to pH 11, but it is the most expensive chelate in this group.

In addition to pH, selecting the right chelate must also take into account the ion being delivered and the plant being fertilized. This is all much more complex than implied by product advertising, especially that geared toward home gardeners. For smaller gardens, it is best to bypass these products.

EDTA is one of the most popular chelates used in agriculture, and unfortunately, it is not quickly decomposed by microbes. It can be toxic to them. As a result, an increasing amount of EDTA is being found in our water systems.

Fertilizer Calculations

When applying fertilizer, you need to know the amount being applied, the area to which it is applied and the amount of nutrients in the fertilizer. The following calculations will work for any units of weight and area (empirical or metric) provided the same units are used throughout the calculations. Don't mix pounds with kilograms, or square-foot with acres without first converting to the same unit of measure.

How much fertilizer is need to provide 1.0 lb. nitrogen per 1,000 sq. ft.?
A fertilizer with an NPK of 26-5-10, contains 26% nitrogen. Divide the weight of nitrogen needed by the % nitrogen in the fertilizer.

1 lb. ÷ 0.26 = 3.8 lb. (Note that when dividing by a %, you use the decimal equivalent.)

Use 3.8 lb. of 26-5-10 fertilizer to apply 1 lb. nitrogen per 1,000 sq. ft.

How much phosphate is applied with 3.8 lb. of 26-5-10 fertilizer on 1,000 sq. ft.?
The fertilizer contains 5% phosphate (P_2O_5). To calculate the amount of phosphate added, multiply the weight added by the percent.

3.8 lb. × 0.05 = 0.19 lb. phosphate.

Applying 3.8 lb. of 26-5-10 fertilizer to 1,000 sq. ft. will add 0.19 lb. phosphate.

How many bags of fertilizer are needed to apply 2 lb. of potash per 1,000 sq. ft.?

To answer this question, you also need to know the amount in a bag, the NPK of the fertilizer, and the total area to be covered. Assume the bags contain 50 pounds, with an NPK of 10-10-10, and that your area is 12,000 sq. ft.

Calculate the total amount of potash needed by multiplying the amount per 1,000 sq. ft. by the total area.

2 lb. × (12,000÷1,000) = 24 lb.

A total amount of 24 lb. potash is needed. A 50 lb. bag of fertilizer contains 10% potash, or (50 × 0.1) 5 lb. potash. Divide the total amount needed by the amount in each bag.

24 lb. ÷ 5 lb. = 4.8 bags

A total of five bags are required to apply the required potash.

Applications of Fertilizer

Fertilizer can be applied at one time of the year or spread over several applications. Since nitrogen moves quickly through soil, use it in smaller amounts throughout the year and only when plants need it. A late fall application of synthetic nitrogen can be a complete waste of resources since it will leach away before plants can use it in spring.

Soil with a higher CEC will hold on to nutrients longer than ones with a low CEC.

Organic Fertilizers

The following are some of the more popular types of organic fertilizers.

Compost

Compost is an ideal slow-release fertilizer. Add a half-inch layer over the whole soil, and let nature take it into the soil. An annual application will produce a healthy soil over time. It can also be incorporated into a new bed when it is prepared.

The NPK value is fairly low, usually in the order of 1-1-1, but the actual values will depend on the input ingredients.

Manures

You can think of manure as being fresh compost. The animal's digestive system has started the decomposition process, but much of the organic material is still intact. Urine is usually mixed in with manure, which provides a quick release source of nitrogen. Urine is also cause for concern because it can have nitrogen levels that are high enough to burn plants. It is best to let manure sit for a while. Poultry manure has some of the highest levels of nitrogen and should be partially composted before it is used. Horse manure tends to be one of the safest to use, but even it should sit for a couple of months before use.

The NPK of fresh manure, approximately 1-1-1, varies depending on the animal source, the feed that was used, and the bedding material. You will find various tables online, but remember that values change over time. NPK values are usually stated on a weight basis, and manure dries over time; old dried manure has higher values than fresh manure that contains a lot of water.

The relative amounts of nutrients also change. The urine portion holds most of the potassium and is easily lost to leaching. About half of the nitrogen is in a form easily lost to the environment, while phosphorus is mostly stable as large molecules. As much as 90% of the nitrogen can be lost in the first three weeks if not properly handled.

One problem with manure is that it tends to contain more phosphorus than plants use, relative to the nitrogen. Over time too much manure can raise P to toxic levels.

Manure is best used by spreading it fresh on top of soil, away from plants, so that it can age in the field.

Rock Dust

Rock dust is a darling of organic and permaculture groups, who highly recommend it. Their logic goes something like this. We continue to use land to grow crops, and each time we harvest them, we

Soil Myth: Soil Is Running Out of Minerals

It seems to make common sense that we are using up the minerals in soil. Certainly various organic and environmental groups point to this as a big problem, but there is little, if any, evidence that this is true.

If soils were running out of minerals, we would need to be adding much more fertilizer. In reality, we add relatively small amounts of fertilizer, mostly to increase yields, not to grow plants.

The food we are producing is at least as nutritious today as years ago, in some cases, more nutritious. That is hard to explain if the soil is unable to supply enough nutrients.

Consider this calculation. Agricultural soil contains 1%–5% iron—let's use an average of 2%, which is 4,500g/sq. m. (0.9 lb./sq. ft.), in the plow layer. A carrot weights 50 g and contains 0.3 mg of iron. So that 1 sq. m. of soil has enough iron to grow 15 million carrots. We have not started to use up the iron in agricultural soil.

There is very little evidence that we need to remineralize our soil.

are removing minerals from the soil. It is only natural that we should replenish these in a process called remineralization.

This is done by adding rock dust on an annual basis. Also known as rock powder and rock flour, it is pulverized rock, which can be human-made or occur naturally. Cutting granite for commercial use produces granite dust. Glaciers naturally produce glacial rock dust. When found near ancient volcanoes, it consists of basalt rock.

To be effective, the rock needs to be ground into a very fine powder, which makes it easier for microbes to decompose. Its NPK value is around 0-0-1, although it does contain many micronutrients.

The problem with rock dust is that there is very little evidence that its minerals are solubilized, except at glacial speeds and then only in very acidic conditions. I have personally asked several manufactures for research to show that rock dust increases the nutrients in soil, but they have been unable to provide any such evidence.

One commercial product makes a big deal about the fact that its product contains 74 minerals, yet plants only use about 20 of these. Until there is clear evidence that rock dust adds nutrients to soil in the short term, there seems to be little point in using it.

Rock Phosphate

Rock phosphate is a type of rock that contains a very high level of phosphorus. The material is mined, ground into a fine powder, and made available to gardeners. It does qualify as an organic fertilizer by the organic certification bodies.

This same material is also used to produce superphosphate and ammonium phosphate, two synthetic fertilizers, which are not allowed by organic certification. That has always seemed very strange to me since both products are processed, have an environmental impact to produce, and neither originates as living organisms—at least not in recent times. And yet one is organic and the other not.

Rock phosphate has the same issues as other rock products. Unless the soil is very acidic (less than 5.5), rock phosphate is not soluble in water, which means the P is not readily released into soil for plants to use. This will only happen after many years, with one source quoting 100 years.

Superphosphate and ammonium phosphate are both very soluble and release phosphate to plants quickly.

Blood Meal

This by-product of the slaughter industry is essentially dried blood. With an NPK of 13-1-0.6, it is one of the highest sources of nitrogen acceptable for certified organic gardening. Unlike other organic material, this one decomposes quickly and releases its nutrients in months, not years. It is a good source of nitrogen, but does not build soil structure as well as other products.

Fad Products

The gardening industry has very few regulations, and producers are free to make almost any claim on their products, which don't even have to work. This has led to the introduction of many fad products.

Vitamin B1, the hot kid on the street for many years, is now known to provide no benefits for plants. Kelp is the new darling, with all kinds of undocumented special benefits ascribed to it. Fish fertilizer, compost tea, milk, and molasses are just some of the other products that have unsubstantiated claims.

Fish fertilizer does contain nutrients and does work as a fertilizer. Compost tea is made from compost, and so it contains nutrients, but claims that the added microbes have a superior effect on soil are not supported by science. Both milk and molasses are made up of organic molecules that decompose into nutrients. None of these products has shown special powers.

Over time, many of these fad products are replaced by new ideas because they don't work as advertised. In actuality, any organic material is essentially the same as any other: over time it decomposes into basic nutrients. There is limited scientific evidence that special compounds in any one of them have significant effects on plant growth or on soil health.

Biostimulants

Biostimulants, a new class of product that is increasingly popular, has been defined by the Eurpean Biostimulants Industry Council as "contain[ing] substance(s) and/or micro-organisms whose function when applied to plants or the rhizosphere is to stimulate natural processes to enhance/benefit nutrient uptake, nutrient efficiency, tolerance to abiotic stress, and crop quality." They include amino acids, peptides, humic acids, seaweed extracts, probiotic bacteria, and special minerals, like cobalt and selenium. Biostimulants are currently a hot topic, despite the fact that they have little scientific support.

CHAPTER 12

Solving Microbe Issues

I am sure that it is now clear to you that soil health is all about healthy microbes. Your goal as a gardener is to have lots of them and to have a very dynamic community. The problem for the gardener, and even the scientist, is that it's almost impossible to identify and quantify them. You never know what you have.

The best you can do is look at the soil and make a guess about the microbes. Low compaction, high organic levels, and good aggregation all indicate a vibrant microbe community. The best you can do is manage the soil and your activities and do those things that improve the microbe community, and then have faith that they will come.

Pathogenic microbes are more evident because they cause visual symptoms on your plants. Once you see those, you can deal with them. To manage your non-parasitic microbes follow these guidelines:

- Assume that you have lots—even poor soils contain lots of microbes.
- Improve the soil to encourage a wider diversity.
- Maintain good levels of organic matter—their main food source.
- Keep soil moist and mitigate extremes of temperature as much as possible.

Inoculation

Numerous companies have developed products to help alleviate the public concern for a healthy microbe population. You simply buy a bottle of microbes, add them to your soil, and they will prosper and make healthy soil. Many of these biostimulants have no scientific support to show that they work.

As a consumer, it is impossible for you to know that the product contains the right microbes or that they are even viable. A recent study of over 20 commercial products found that some never included the advertised microbe, as verified by DNA analysis, and in many cases, the number of live microbes was far below the advertised amount.

Even if your soil is missing some varieties of microbes, adding them will work only if they want to live in your soil and are able to compete with the current inhabitants. Remember that the amount of microbes already in soil is huge compared to the small amount you might add. Most inoculants do not survive very long.

Rhizobiam for Legumes

Rhizobiam are known bacteria that form a symbiotic relationship with legumes to form nitrogen-fixing nodules. The species and the mechanism involved are fairly well understood, and they have been proven to work.

They can be purchased as an inoculant that is used to coat seeds before planting. Once in the soil, the bacteria grow and form associations with the seedling. This is recommended in any soil that has not grown the legume in the past four years. It is also important to match the species of rhizobium to the legume.

Mycorrhizal Fungi

Mycorrhizal fungi certainly play a key role in nature and create beneficial associations with most plants. They have also been shown to have benefits for plants grown in soilless mixes and in greenhouses.

Fungal inoculation of soybeans, with AM mycorrhiza, produces larger beans in low-phosphorus soils but has no benefit in soils with adequate phosphorus levels. Such inoculation also has been shown to increase survival rates in young trees during a reforestation process. Added to newly planted grapes, it increased growth and yield in the first year. It is important to note that, in each case, they used specific species that had been previously tested and verified to work with the host plant. Proper host-fungus pairing is critical.

There are also studies that show that mycorrhizal fungi can result in less plant growth and an increased incidence of nematode infection. We still have much to learn about these fungi.

There is limited evidence of any benefit for adding them to ornamental plants or garden soil. Fungi spores are ubiquitous and are already in most soils. The very poor subsoil that remains after some building projects may be devoid of them, but inoculating that soil won't work since the soil is not suitable for their growth. You first have to fix the soil problem, and once you do that, the fungi will find it naturally.

Probiotics

Probiotics for soil is the same idea as probiotics for your intestines. They are a combination of microbes, usually bacteria, that you buy and add to your soil. Much of the hype around these products discusses the benefit of our own gut bacteria and then infers that adding bacteria to soil will benefit plants in a similar way. The two systems are completely different.

Very limited work has been done to show positive results for probiotics.

Solarization

Pathogens can be controlled by solarization, whereby plastic is laid on the soil, and the heat of the sun is used to kill them. The soil temperature in hot climates with lots of sunlight can reach 122°F (50°C) in a few hours, down to a depth of 4 inches (10 cm). The average increase is about 18°–27°F (10°–15°C). This procedure does work in colder climates, but the treatment period needs to be extended.

Wetting the soil activates fungi, seeds, and bacteria, making them more susceptible to heat. The water also helps with heat transfer into the soil. Fungi in wet soil are killed after a few hours at 111°F (44°C) and after only a few minutes at 122°F (50°C).

Some heat-tolerant fungi can survive these temperatures, but they tend not to be pathogenic. Solarization changes the microbe population in favour of organisms that do not cause diseases, and the surviving organisms are better able to compete with newly introduced pathogens, but the effects last only a few months.

Placing cabbage plant waste material under the plastic produces toxic chemicals like aldehydes and isothiocyanates, which can increase the death rate of pathogens. The problem with solarization is that it also destroys beneficial microbes.

Controlling Pathogens

Tilling can be used to control surface microbes that attack the crown of plants. In some cases, burning the pathogens will kill them and control their population, but it also has the potential of spreading them around. Non-tilling increases biodiversity and can also be an effective control mechanism. Planting in cold soil can also increase disease pressures.

Seed is more prone to disease at planting time, especially in wet conditions. Wheat seed sown in wet soil is more likely to be infected by pythium. Drip irrigation that is too near plants can cause problems since nematode populations are higher near the dripper. Moving the dripper or burying it to allow the surface to dry out will improve the situation. In other cases, higher levels of water encourage antagonistic microflora to grow and compete with pathogens.

Pathogen control is a complex subject, and in most cases, each pathogen will require a different solution. Identify the problem organism, and then search for a solution. The following can be helpful in controlling pathogens:

- Use disease-free seeds and plants.
- Use certification programs such as certified potatoes and garlic.
- Obey quarantines.
- Control pH.

- Rotate crops.
- Use organic matter, which encourages decomposers that attack parasites.
- Use plants that supress some organisms, e.g., marigolds for root-knot nematodes.
- Select disease-resistant cultivars.

Compost Tea

Compost tea, also known as manure tea, has become very popular as a soil amendment. There are many recipes, but everyone talks about it as if this is a single type of amendment. That is one reason it is very difficult to figure out what it actually does, or whether it has any effect at all.

The ingredients can be just about anything, including dead weeds, grass clippings, manure, and compost. It can be brewed in various ways. Some people just add water and let it sit for a few days. This usually results in anaerobic decomposition and really stinks. Others feel that oxygen should be provided by bubbling air through the mixture during the brewing process, a method called AACT (actively aerated compost tea).

The benefits ascribed to compost tea include the following.

- Increased nutrient levels
- Decrease diseases
- Increased soil microbes

Increased Nutrient Levels

The nutrients in compost tea will depend very much on the input ingredients. Some decomposition will take place during the brewing process. This will release some of the nutrient ions that are trapped inside cells and add them to the tea, but most are tied up in large molecules, and a couple of days of brewing does not significantly change that.

The brewing process can't increase the nutrients—it is physically impossible. The total nutrients in the brew will be the same as those found in the input ingredients. Studies that show increased growth from tea usually compare tea to water. It is no surprise that compost or manure has more nutrients than water.

Brewing compost tea might speed up the release of plant-available nutrients, although I doubt this is significant, but it can't increase the total amount.

Decrease Diseases

The usual claim involves spraying the compost tea right on plants, where the microbes in the tea help decrease diseases. Some studies show positive results, and many show negative results. One problem with much of this work is a lack of control over the input ingredients. Most of these studies don't identify the microbes in the tea, making it impossible to compare brews or studies.

For this to work, the sprayed-on microbes would need to colonize the surface of leaves, which requires that the leaves have the right habitat for them. Microbes from anaerobic teas will likely die when exposed to air. Leaves are covered with millions of microbes per square inch, and competition is fierce. It is more likely that native microbes will outcompete any that are sprayed on.

Compost teas might suppress certain diseases, but the science does not yet support this.

Increased Soil Microbes

Since soil microbes are critical for healthy soil, everybody assumes that adding more will improve soil. Brewing compost tea does increase both the fungal and bacterial populations. The problem is that it is near impossible to know which species are growing, and it is certainly possible that the brew contains pathogens like E. coli.

Even if the microbes are good microbes, the number in the tea is miniscule compared to those in soil. When they are added to soil, they need to first survive the new environment and then compete with the residents that are already there. More than that, they need to prosper so their numbers increase enough to become a major player in the new population.

The scientific evidence does not support the idea that added microbes will survive. In really poor soil, the environment is not suitable for them, and in better soil the competition is fierce. It is quite possible that, with more research, certain species will be identified that are beneficial, but we are not there yet.

Best Practice for Increasing Microbe Populations

Except in specific well-documented situations, the best way to increase microbe populations is to improve their environment so that the existing microbes grow better. If you build healthy soil, they will come. As the quality of soil improves, the microbes will find it and will prosper. All of the following improves their environment:

- Supply fresh organic matter.
- Provide good aeration by reducing compaction and increasing aggregation.
- Maintain even moisture levels.
- Keep temperatures between 77°F and 99°F (25°C and 37°C) for as long as possible by using mulch.
- Raise pH in very acidic soil.
- Provide adequate nutrients, but not excess amounts.
- Limit chemical treatments for pathogenic microbes. Anything that kills pathogens also kills beneficials.

CHAPTER 13

Increasing Organic Matter

The degradation of today's farmlands can be directly attributed to a loss of organic matter. Both the techniques used to work the land and the reliance on synthetic fertilizer have produced soil with a low level of organic matter. This is the land that becomes the starting point for many new homeowners and micro-farmers. Increasing the level of organic matter is the most import step toward creating healthy soil. Unfortunately, there are no fast solutions.

Options for Adding Organic Matter

Which is the best organic matter? There is no single best material, and the best one for you depends on your location, but these guidelines will help select a suitable product.

- Cheap is good. A low price usually means that the material has undergone limited processing and has limited transportation costs, both of which are good for the environment. The best-priced items are usually waste products from some industry.
- Use locally sourced material.
- Material with less decomposition is better since it provides a longer feed for the garden.
- A C:N ratio of 30:1 is ideal, but extra nitrogen can be added to material with a high carbon value.
- Neutral pH is best.

Before selecting organic matter, understand your goals. Is there a specific problem you are trying to solve? For example, wood ash is high in potassium and would be a good selection if potassium is deficient. However, it is also quite alkaline, so it is not a good choice for high pH soils.

How quickly do you need to make nutrients available to plants? Most organic matter does not feed plants quickly, but some things like blood meal do. If a quick feed is required, the best solution might be a combination of synthetic fertilizer for immediate nutrients and organic matter for longer-term feeding.

Other options that decompose more quickly include fish powder, kelp powder, and liquid kelp. Options that are very slow to release nutrients are most of the minerals, such as rock phosphate and greensand.

Salinity can be a problem with some organic material, especially manures, mushroom compost, and biosolids. Suitability of these products depends on their salinity value and how much you use. Using small amounts is usually not a problem. If you use significant amounts, have the material and/or soil tested.

Variety is a good thing. It is a good idea to vary the materials used to ensure that one or more nutrients are not out of balance. It might also increase the diversity of microbes.

Compost

Compost is one of the best options for adding OM, especially if you make it yourself. The input material will vary, but you have full control over the process. Nutrient levels are usually low, in the order of 1-1-1, but the material is excellent for building soil structure. Most home-made compost does not get hot enough to kill seeds and pathogens. Municipal compost is made at much higher temperatures that take care of seeds and most pathogens. It is normally tested for heavy metals and certified safe. In some jurisdictions, it may contain herbicides, which were used on the input material, and it may contain small bits of plastic.

Compost made mainly from manure can contain high salt levels.

The differences between compost made from horse, cow, and sheep manure are insignificant.

Mushroom Compost

Mushroom compost is the material left over after growing mushrooms. Growers start by making a compost from all kinds of material, including bedding straw, manures, cottonseed meal, soybean meal, potash, gypsum, lime, and other fertilizers. This is left to compost for a few weeks before being used to grow mushrooms. At the end of the harvest, the material is still quite nutritious, so it is made available to gardeners.

It goes by various names, including mushroom fertilizer, spent mushroom substrate (SMS), and spent mushroom compost (SMC). The contents vary depending on the grower and how it was treated after use. In some cases, it is quite fresh, and in others, it has been left for a few months to age.

There is concern that mushroom compost contains too many salts. This idea probably originated as a result of trying to use it as a growth media in pots. The salt levels were just too high for this. When small amounts of the material are used to amend soil, the salts are not a problem. Aged material has lower salts and may be a better option if you are using large amounts.

It is also necessary to understand which salts it contains. Sodium levels are low. It is the nutrient ion levels that are high, and these are what plants need.

Mushroom compost is excellent for adding organic matter and nutrients in the short term.

Manure

Fresh manure is readily available at a low price and is excellent for building soil structure. Very fresh manure has a high salt content, and the urine in manure can have high nitrogen levels. This is especially true for chicken manure. Most people will let fresh manure sit for several months so that excess salts can leach out. A better approach is to use it as mulch, keeping it some distance from plants. In this

Percent Nutrients in Fresh Manure.

Manure	Nitrogen	Phosphate	Potash	Sulfur	Calcium	Magnesium
Dairy cattle	.5	.2	.4	.05	.3	.1
Beef cattle	.55	.4	.5	.05	.15	.1
Poultry	1.15	.55	.5	.15	1.8	.3
Swine	.5	.15	.4	.15	.5	.1
Sheep	1.4	.2	1.0	.1	.5	.2
Horse	.65	.25	.65	—	—	—

way, the leaching nutrients are captured in the soil and benefit the plants.

Horse manure contains a lot of bedding material and has lower salt levels, allowing you to use it much sooner.

Weed seeds can be a problem depending on the source of material.

Leaf Mold

Fall leaves have a high C:N ratio, and on their own they decompose very slowly, forming a partially decomposed substance called leaf mold. It makes a good soil amendment or mulch.

Straw and Hay

Straw and hay are readily available in many areas and make excellent mulch in a vegetable garden. This material has a high C:N ratio, but that's not a problem when used as a mulch. It is not suitable as an amendment unless composted first. There is some concern about residual herbicides, which can inhibit seed germination.

The material decomposes slowly and needs to be replaced about every year or two depending on climate. Hay is reported to have more weed seeds, but it is also higher in nitrogen. When it is used as a mulch 4 to 6 inches deep (10 to 15 cm), weed seeds are not really a problem.

Wood Products

Waste wood products can be anything from wood chips to sawdust. This material is low in nutrients and has a very high C:N ratio, which causes it to decompose very slowly. Because of the low nitrogen levels, it should not be incorporated into soil, unless it's combined with another nitrogen source. It is best used as mulch. Sawdust can be used as a mulch, but it is really too fine and may restrict water from percolating through. Neither wood chips nor saw dust needs to be aged if it is used as a mulch.

As mulch, the material lasts a long time and provides a very slow feed. Its main benefit is the addition of carbon to the soil.

Plant-based Meals

There are several plant-based meals that are popular as soil amendments, including soybean meal, cotton meal, and alfalfa meal, which all have good nutrient levels. Generally sold as finely ground material, they are easy to spread and apply to soil.

They contain higher amounts of nitrogen than some other products and may inhibit seed germination. The meals are highly processed and as such decompose faster than other organic material, providing nutrients for about one year. Some consider them to be more of a fertilizer than a source of organic material.

They are not great for building soil structure, but they are weed free.

Agricultural By-products

A number of agricultural by-products are good options, including ground corncobs, apple pomace, rice hulls, cocoa bean hulls, and peanut hulls. The courser material works well as mulch; finer material can be used as mulch or be incorporated into soil.

Check on the C:N ratio of the product you use, and if the ratio is high, add additional nitrogen when using them.

Peat Moss

Peat moss is partially decomposed plant material. It forms naturally in wet anaerobic conditions that are usually very acidic. Sphagnum peat moss develops from various sphagnum mosses, but peat can also form from other plants.

Soil Myth: Peat Moss Acidifies Soil

Because peat moss is quite acidic, many recommend it for acidifying soil. When added to alkaline soil, the buffer capacity of the soil quickly neutralizes the acidity of the peat moss, and within a week, the soil pH is back to normal. Peat moss does not work for acidifying soil.

The material is mined from peat bogs, which is quite a destructive process. Consequently, many now believe harvesting it is not ecofriendly. The reality is that our natural peat reserves increase a hundred times faster than the rate at which we use it. The world is not running out of peat reserves, and alternatives, such as coir, are not any more ecofriendly.

Using peat for starting seeds and in containers is an environmentally acceptable use of the product. Using it to add organic material to larger gardens and farms is not a sound practice, although it does work well. It is better to use other waste products.

Peat moss does add organic matter, is very acidic, and contains almost no nutrients. The material is light and easily blows away in wind, making it a poor choice for mulch.

Coir

Coir, a by-product of the coconut industry, consists of chopped up husk. It holds a lot of water and makes a good amendment.

The issue with this product is environmental. In its natural form, it contains high levels of sodium, and these need to be washed out during production of the coir, requiring large amounts of fresh water. This processing is done mostly in India and Shri Lanka, where it is causing significant environmental damage to local fresh-water resources. It then must be shipped a long way to get to your garden. Although using a waste product sounds environmentally sound, in this case it's not.

Cover Crops

Cover crops, also called green manure, are planted to provide fresh organic matter and nutrients to soil. In covering the soil, they reduce damage caused from rain, prevent weed seed germination, and decrease erosion. Consider planting cover crops any time the soil is not being used for crops.

A wide range of plants can be used as a cover crop. They generally germinate easily, require little extra maintenance or fertilizer, and grow quickly. Legumes are a good option, providing additional nitrogen to the soil. Others are selected because the cold winter kills them, effectively removing them before spring crops are planted.

Some well-known examples of allelopathic cover crops are rye, hairy vetch, red clover, sorghum Sudan grass, and some mustard family species. In one study, rye cover crop residues were found to have provided between 80% and 95% control of early season broadleaf weeds when used as mulch during the production of different cash crops such as soybeans, tobacco, corn, and sunflowers.

When selecting the cover crop, consider how you will end the crop. Annual types may be killed off by winter, but then they are not growing in early spring. Others will be growing in spring and require either chemical or mechanical methods to kill them before planting the main crop. In most cases, it is also important to ensure that the cover crop does not go to seed and create a weed problem.

Before you select a cover crop, understand your main goals. Are you trying to add nitrogen or produce as much organic matter as possible? Does it need to control any special pest issues?

Your local environment is also critical. In cold climates, the growth period for a cover crop can be short, which limits your choices. In warmer climates, the cover crop may be in the ground longer than the main crop.

The method of harvesting is also key. Some annual cover crops grow well in fall and then are killed over winter, so harvesting them is not an issue. Perennials need to be dealt with before the main crop is planted, or else they become weeds and compete with the main crop.

Pros and Cons of Various Cover Crops.

Cover crop	Type	Comment
Crimson clover	Winter annual legume	Good choice for warmer climates, adds significant nitrogen
Field pea	Winter annual legume	Establish quickly, grows cool, significant residue
Hairy vetch	Winter annual legume	Can take freezing, produces significant vegetation, decomposes quickly
Berseem clover	Summer annual legume	Establishes easily, develops a dense cover, drought tolerant
Cowpea	Summer annual legume	Grows well in hot climates, deep-rooted, drought tolerant
Velvet bean	Summer annual legume	Climbing vine, aggressive, produces a thick layer
Alfalfa	Perennial legume	Needs well-drained soil, likes high fertility
Crown vetch	Perennial legume	Needs well-drained soil, establishes slowly
Red clover	Perennial legume	Vigorous, winter-hardy, establishes easily
Sweet clover	Perennial legume	Vigorous, grows in compacted soil, drought tolerant
White clover	Perennial legume	Limited growth, stays short, tolerates shading
Cereals	Grass	Nutrient scavengers, extensive root systems, large amounts of residue, high C:N ratio
Winter rye	Grass	Very winter hardy, establishes easily, residue supresses weeds
Oats	Grass	Not winter hardy, stops winter erosion
Annual ryegrass	Grass, annual	Grows well in fall, extensive root system, significant growth
Sudan grass	Grass, summer annual	Vigorous grower, suppresses weeds
Buckwheat	Summer annual	Killed by frost, suitable for low fertility, grows quickly
Brassica family	Annual/biennial	Large root system, good in compacted soil, extracts high level of nutrients

Vermicompost

Vermicompost, the material left behind after worms eat and digest kitchen scraps, manure, and other types of organic matter, is aged worm poop, also referred to as worm castings. It contains lots of bacteria and partially decomposed organic matter.

Commercial vermicompost has an NPK value of around 2-4-1, which is much more phosphorus than normal compost. Care needs to be taken when using this product to ensure that phosphorus levels in soil do not get too high.

Micro-farms can set up their own vermicompost production. The input ingredients can come from the farm or other local sources, and management of the worms is relatively easy, particularly in warmer climates.

Bokashi Compost

In North America and Europe, bokashi is a process used in homes to get rid of kitchen scraps. In much of the rest of the world, especially Asia, bokashi is carried out on a much larger scale. Although it's called composting, it is really a fermentation process. Organic input ingredients are combined with special bokashi starter mixes called Effective Microbes (EM) and allowed to ferment before being added to compost piles or directly to soil. The liquid created during the process, called bokashi tea, is used as a fertilizer.

Bokashi produces less CO_2 during the fermentation process than traditional composting, making it better for the environment, and it adds more carbon to the soil. In traditional composting, a lot of nitrogen gets converted to atmospheric nitrogen and is lost. Bokashi requires very little nitrogen, and more ends up in the soil. The pH of bokashi ferment is about 4.

It is not clear what happens to the bokashi ferment once it is added to soil. Does it decompose as slowly as compost, or are nutrients released much more quickly? It is odd that a process that is used so much is not better understood.

Biochar

Biochar is being promoted as a special kind of charcoal that has many benefits as a soil amendment, while at the same time being eco-friendly and sequestering carbon for many thousands of years. This sounds like a perfect product.

The claims made for biochar include that it
- Increases yield
- Increases fertilizer efficiency
- Removes pollutants and pesticides
- Mitigates climate change
- Increases soil moisture
- Increases soil pH
- Increases soil microbe populations
- Increases cation exchange of soil

Most of the studies that make these claims have been conducted in the lab, using very artificial systems. Very few field tests have been done, so we don't know what biochar does in real soil.

Both charcoal and biochar absorb chemicals. They do have negatively charged sites that increase CEC (cation exchange capacity), but if they absorb pesticides or nutrients, then this can be detrimental to plants since it makes these products less effective. We also know that activated charcoal, which is used to scrub chemicals, gets saturated and needs to be replaced. I would expect that biochar in soil will also become less effective over time.

Biochar does increase the pH of soil and holds more water, neither of which is a benefit in alkaline clay. It does increase microbe populations.

A current study that reviewed the existing literature concluded that half of the biochar studies reported an increase in plant yield, while 20% found a decrease, and 30% reported no change.

What is the difference between charcoal and biochar? That is not really clear. Some sources say biochar has to be made at high temperatures, but others disagree. The UC Davis biochar database says it is charcoal that is primarily used for soil amendment and not for heating. Other sources agree that it is defined by how it's used and not by the method used to manufacture it.

What is clear is that manufactured products vary greatly, which makes it very hard for consumers to know they are buying a product that works. At the present time, biochar is a fad product, and I would not recommend adding it to soil.

Biosolids

Biosolids, a by-product of sewage treatment, consist of processed sewage sludge. It may be found alone or composted with leaves or other organic materials. Developed countries produce a lot of this material, but there is a reluctance to use it.

The main concerns for biosolids are heavy metals, pathogens, and salts. In an effort to make this material more acceptable, it is now tested and classified. Class A biosolids are tested to be free of fecal coliform, salmonella, and heavy metals, but have limited testing for chemical contaminants. They are approved by the US EPA for use on farms and home gardens. European regulations are tighter.

Numerous studies have shown that food produced using biosolids is safe to eat, but it is not allowed on organic certified farms. Before you use this material be sure to check the salt levels.

CHAPTER 14

Dealing with Structural Problems

This section will discuss several structural problems, including compaction, drainage issues, modifying soil texture.

Compaction

The best way to manage compaction is to prevent it in the first place. Reduce or eliminate anything that puts pressure on the soil. Dealing with an existing compaction problem can be difficult, and the best solution will depend on the status of the site. A new site or an area used for annual crops can be dealt with differently than an established landscape.

Natural methods are effective in the top foot of soil. If compaction extends past that point, or if the compaction is so severe that plants and soil microbes can no longer thrive, natural processes won't work and mechanical means should be considered.

Subsoiling

The goal of subsoiling, also called soil ribbing, is to fracture the soil with minimal disturbance to plant life and topsoil. This is done with special equipment called a subsoiler or ripper that has deep cultivator-like teeth that disturb the soil as it is pulled along the field. It is common to incorporate compost at the same time as subsoiling.

The effectiveness of this process depends on many factors, including various soil characteristics, the type of equipment used, and the speed of the plowing. Each situation has to be assessed differently and requires an expert in the field.

Air Spading
Air spading uses compressed air to break up compacted soil, especially around existing trees. It is quicker than conventional digging and does much less damage to tree roots. The technique is most suitable for a small area.

Rototilling and Tilling
Tilling can be effective in reducing compaction in a new area, provided the compaction is not too deep. Organic matter or sand can also be incorporated at the same time.

Core Aeration
Core aeration, which removes small vertical cores of soil, about 3 inches deep, is a common practice for lawns. The holes can then be filled with organic matter or left alone.

Amending with Inorganic Material
Existing poor soil can be amended with material like sand to decrease its density. Research at Cornell has shown that 70% by volume would need to be added to have a significant effect. The material needs to have a uniform, medium to large, size. This prevents small particles from filling pores.

Amending with Organic Material
Organic material can also be used for amending soil, but this should be done over a larger area, and not just in planting holes. To effectively reduce compaction, it is necessary to add at least 25%, by volume, to a depth of 18".

Mulching
Mulching soil with organic matter may be a slow method for improving compaction, but it is very effective and is the best option for established plantings.

Drainage Issues

The solution to any drainage issue starts with an understanding of the cause. Watch the movement of water during and after a heavy rain. If the water runs in the wrong direction, can something be done to divert water away from the problem area? If this is not practicable, consider ways in which water can be removed from the problem area.

Flooding occurs most often on clay soil and compacted soil because water percolates through such soil very slowly. Instead of being absorbed, water runs off and collects in low-lying areas. Improving the soil increases infiltration and reduces the runoff problem. Mulching also slows down the movement of water across the surface of the soil, giving it more time to absorb.

Proper Grading

Check the grading by actually measuring it. The human eye is very poor at seeing slopes, so it is a good idea to actually measure the slope with a level. Changing the slope to divert water away from problem areas can solve many drainage problems. In some cases, small berms might be needed.

Drainage Tiles

To move water away, install subsurface tile drains, perforated plastic tubes that are buried 12 to 18 inches (30 to 45 cm) beneath the soil surface. Water filters down through the soil and collects in the tubes. They are laid on a slant so that water runs away, usually emptying into a drainage ditch or pond.

Many farms in Ontario and the US Midwest have this system installed to drain spring rains and allow earlier planting. When used around homes, they are called French drains.

Surface Drains

Where water collects in a low spot, you can install a surface drain, which is very similar to the grates found on paved roads. It is connected to underground drainage tiles that remove water from the area. The advantage of a surface drain is that the water does not have to seep through soil before it is removed, and therefore it works very well in heavy or compacted soil.

Dry Wells

A dry well can work in small areas that collect water. The hole, which is filled with rocks or plastic containers designed for the purpose and then covered with soil, creates a large air space underground. Excess water on the surface of the soil drains into the air pocket and keeps the surface dry.

Raised Beds

One of the benefits of raised beds is that they drain better because of their height and also because their soil is usually amended to be more porous. This works well for plants, but they tend to funnel the water into the walkways between the beds. Adding drainage tiles under the walkways can help.

Changing Hardpan

Hardpan is a condition in which a hard layer of soil, or rock, exists below the surface of the soil. It prevents water from draining any farther, causing the soil above it to stay wet. Because the cause and solution can vary widely, it is best to bring in an expert to assess the situation.

If the hardpan is deep enough, drainage tiles can be installed above it. An alternative solution is to either remove the hardpan or at least create cracks in it. In some cases, this can be done with deep plowing or with a broadfork. More difficult hardpans may need to be drilled.

In some cases, a hardpan can be resolved slowly using organic matter, adjusting acidic soil pH with lime, or by adding gypsum.

High Water Table

As water percolates through soil, it reaches a point where it pools, effectively creating an underground pond. A high water table exists when the surface of this pond is close to the surface of the soil. The height of the water usually changes with seasons; sometimes it is a problem only in spring as excess rain and snow melt flows underground.

On hilly property, the water table is usually a problem only in the low-lying areas.

There are only two solutions to a high water table: build the soil up or remove water. Removing hardpan and adding drainage tiles may lower the level enough so that crops can be grown. Adding soil can be expensive but is effective.

You must first determine the level of the water during various seasons to figure out the best solution. A high water table in midsummer can actually be an advantage, provided the level is low enough to allow crop roots to grow.

Modifying Soil Texture

Ideal soil has 40% sand, 40% silt, and 20% clay, but almost nobody has this type of soil. Changing these ratios is not practicable because you would need very large amounts of material and some way to blend them together. A better option is to grow plants that like your soil. Adding organic matter will help both ends of the extreme.

Clay Soils

Fine textured clay soil is difficult to work, especially when it is too wet or too dry. When dry, it forms vary hard lumps; when wet, it is sticky and remains in big chunks. Making a seed bed for small seeds can be a real challenge. On the positive side, clay holds a lot of moisture and nutrients.

In a new planting area, it is best to amend the whole area before planting. Don't try to amend single planting holes because this creates what is, in effect, a small container in the clay soil. Add several inches of organic matter over the whole area and work it into the top 6 inches (15 cm) of soil. In very heavy clay, it will be almost impossible to mix it well, but even if it gets below the surface, it will start having an effect.

Because existing landscapes can't be amended easily, it is much better to cover the area with organic material as mulch. Compost and manure would be the best choices since these will be incorporated into the soil more quickly, but they tend to allow weeds to grow. Wood chips are much better at preventing weeds, but they are incorporated into soil very slowly and take years to improve the soil.

Addition of Sand

It is commonly believed that adding sand to clay will create concrete. That won't happen because concrete requires cement. What people are trying to say is that the soil becomes harder to work after adding sand. Although there are numerous anecdotal reports of this, no studies confirm it. The results might depend on which combination of the many types of clay and sand you use.

In the UK, Europe, and eastern US, it is common to add sand to clay soil, and there seems to be no issue with the soil getting hard. I have used it in three different gardens with great success. Most of the reports of negative effects come from the western US states.

The soil texture triangle covers all possible combinations of clay and sand, and nowhere is there a type of soil that is hard as concrete.

In very heavy clay, which can only be broken down into smaller clots, the sand coats the clots and makes future digging easier. Course sand of mostly uniform size works best. It won't help improve structure or fertility, but a few inches of sand can make the soil much easier to work.

Addition of Gypsum

Adding gypsum (calcium sulfate) to clay soil is common advice, claiming that it breaks up the clay into more manageable soil. It is also the main ingredient in so-called clay-buster products. This is mostly a myth.

Sodic soil contains a lot of sodium ions, which have the effect of destroying clay structure. The calcium in gypsum replaces sodium ions that get leached out of the soil, thereby improving the clay's structure. It can also help reduce toxic levels of aluminum. Gypsum has almost no effect on soil that has a normal level of sodium.

Gypsum can have some of the following detrimental effects:
• Increases the leaching of iron and manganese

- Causes magnesium deficiencies in acidic soil
- Inhibits mycorrhizal fungi

Be very careful using gypsum; apply it only according to recommendations resulting from a soil test.

Sandy Soils

Sandy soil suffers from three problems. The lack of small pore spaces means that water runs away quickly and plants dry easily. The sand has a very low CEC, so nutrients don't stick to it, and added fertilizer is quickly washed away. Microbe populations are very low because there is no food source or water for them.

The obvious solution to these problems is to add clay, but this is not a practical solution because of the large amounts that would be needed. It is also difficult to mix the clay and sand together to create a good structure.

All three problems are corrected by increasing the organic matter in the soil. Add 4 inches of compost to a new bed and incorporate it into the soil. Mulch existing beds with easily decomposable material like compost and manure, instead of woodchips which take longer to degrade.

Biochar is another option for sand. It holds water and has a high CEC for holding nutrients. This product has the advantage that it degrades very slowly compared to other forms of organic matter, providing long-term benefits to the soil. If you do use biochar, combine it with other types of organic matter that slowly decompose and feed plants and microbes.

A Personalized Plan
for Healthy Soil

CHAPTER 15

Developing a Plan for Soil Health Improvement

The previous chapters have provided extensive information about soil and possible problems that you might have, which is great, but what do you do now? The next step is to put together a plan that will allow you to slowly improve the health of your soil.

This section is designed to guide you through the process of creating a personalized improvement plan for your site. Chapter 17, Soil Health Assessment, will help you gather information about the current condition of your soil. Chapter 18, Soil Health Action Plan, will help you set goals and decide on specific action items for moving forward.

This is very much an iterative process that takes several years, or even a lifetime because changing some characteristics of soil is a slow process. For example, building aggregation takes years, and you can't really rush it. It is also a slow process because most people can't do everything at once, due to time or financial constraints.

You should perform the soil health assessment annually, so that you can track improvements from year to year. It will show you the areas that are improving and ones that still need some work. After each assessment, review the action items and make any changes required.

The forms used in this process are provided in both a printed format and an electronic format. You are free to photocopy them out of the book, fill them in, and store them as you proceed through the process. Alternatively, you can download the files from: gardenfundamentals.com/soil-book-forms/.

The benefit of using the downloadable forms is twofold: you can modify them to suit your requirements, and they can be used electronically, reducing paper waste.

CHAPTER 16

How Detailed Should You Get?

Before you develop a soil health improvement plan, it is of value to assess your situation and decide how detailed you want to get. Most plants will grow in most soils, and for many gardeners, a simple improvement plant will be beneficial and meet all of your needs. A larger garden or market garden may require a more detailed plan.

Also ask yourself how long you plan to garden in your current location. If you move frequently, it may not be worthwhile investing heavily in your soil, although the next owner will certainly appreciate it.

If you are creating a landscape garden, maximum growth is probably not what you need. In fact, many landscape plants do much better on lean soil rather than a high-nutrient one. You want to grow things lean and mean, in which case, you might not want to spend very much time on your improvement plan.

Food-production gardens have different requirements. In many climates, you have a limited time to grow the crop, and you want plants to grow better than average so that you get a good yield. In this situation, it is worth your time and money to develop a more detailed plan and to follow through on improvements.

For a market garden, where you are making a living growing crops, it becomes even more critical to maximize your space. Investments in soil improvement can result in significant profit gains by allowing you to grow plants closer together.

CHAPTER 17

Soil Health Assessment

The best time to start this process is early spring, but it can be done at any time. Step one is to understand your current situation. What are the properties of your soil today? Do you have drainage issues or nutrient issues? How significant are they? You can develop an improvement plan only if you know your starting point.

Use the Soil Health Assessment Form, in Appendix A, to document your current situation. Ideally you should answer all of the questions, but you might already have some good knowledge about the soil conditions in your area and can skip some. For example, many soils are not sodic, and therefore there is little value in getting the sodium or SAR test done.

Some of the soil characteristics on the form can be measured by different tests, and not everyone agrees on which is best or even how they should be done. I have selected most soil tests that are easy to perform by the average gardener with limited experience and relatively no equipment. If you prefer to do the test differently, go right ahead and change the form to suit your methods.

Most of the soil tests have been described in more detail in the relevant sections of this book, which are referenced by page number on the form.

Each test has space for three values: current value, previous value, and a rating. The current value is the result at the time you complete the form, which is dated at the top. The previous value will not be known if this is your first time using the form, but in subsequent years, it will be the value from the previous assessment. This allows you to compare values over time. The rating is a simple numbering system to categorize the current results for each test: good = 1, OK = 2, and needs improvement = 3. It gives you a quick way to identify problem areas.

Chemical Tests

There are a couple of options for doing the chemical tests. You can purchase various kits and meters for doing the tests yourself, but be aware that the cheap ones, sold at garden centers, are not very good, often giving the wrong results.

It is much better to get samples tested by a professional lab. Research the availability of labs in your area and try to select one that is familiar with your soil type. They will use different procedures depending on the soil they test. It is better to use a local lab that is familiar with your soil type than one farther away that may perform the test differently.

You should use the same lab each year. It can be difficult to compare year-to-year test results from different labs.

Soil Sampling Instructions

Most labs will provide you with instructions for sampling your soil, and it a good idea to follow them. If they don't, use the following general instructions.

Determine how many soil samples you will be submitting to the lab. For most gardens, one is enough, but there are good reasons for submitting more. Ask yourself, how are you going to respond to the test results. If you submit two samples, are you going to treat each area differently? If not, there is little point in sending in more than one sample.

Here are some reasons why you might want to submit more than one sample:

- Different areas of the garden have been treated differently in the past.

- The soil texture is clearly different in various sections.
- You are using raised beds, and each one has been customized for a particular crop.
- You have a problem area where plants don't grow well.
- You want to analyse both the topsoil and the subsoil.

The best time to sample soil is spring, with fall being a good second option. Samples should be taken before you fertilize or lime the soil. Use a clean container. To get a good average result for the whole area, you must take several subsamples and mix them together. Five to ten subsamples are suitable for most home gardens; 15 can be used for larger areas. Take the same amount of soil from each spot and mix them well, before taking a final sample for analysis.

If the ground is covered with mulch, plant residue, or live plants, remove them first. Sample just the soil. Remove any stones, plant roots, and large pieces of dead plant material, including wood.

Try to take a core of soil that includes the top part as well as the soil under it, down to a depth of about 6". If it contains a lot of moisture, dry the sample before packaging it. This is easily done by spreading it out on a newspaper overnight.

CHAPTER 18

Soil Health Action Plan

The Soil Health Assessment Form provides you with a snapshot of your current soil health. Items with a rating of 3 are definite problem areas. A rating of 2 indicates that improvement would result in better soil, but it is not an area of concern. Items with a rating of 1 don't need to be improved.

Some of the issues can be improved fairly quickly. If the soil is low in a specific nutrient, you can add that nutrient and fix the problem. You may need to do this annually, but at least it is a quick fix. Other issues on the list, like low organic matter or a small number of earthworms, can't be fixed quickly. There are things you can do now to start the process, but it will take several years of work to convert a 3 rating into a 1.

It is a good idea to take a double-pronged approach by tackling some easy fixes as well as some long-term fixes. Start items that can be fixed quickly right away since they will give you immediate improvements. Problems with long-term solutions also should be started right away since they take a long time. The sooner you start, the sooner these will improve.

Your next step is to decide how many of the 3-rated problems you can tackle immediately. You do not have to work on all of them at the same time. Evaluate both your time and financial situation and try to determine how many you can reasonable tackle right now.

The following table provides some information about how quickly you can change soil characteristics. Set realistic goals for your action plan.

Changing Soil Characteristics.

Difficult to change	Relatively easy to change	Can be changed slowly
Texture	Structure	Level of organic matter
Topography	Drainage	Compaction
Climate	Fertility	Biological activity
	Acidity	Water retention

Soil Health Assessment

Each problem on the assessment form is listed in the following section, along with possible action items to solve it. They are organized in the same order as the Soil Health Assessment Form.

These relatively simple recommendations are focused on the most likely cause of the problem. Every situation is different, and your specific solution may be more complex. Take the time to understand the problem, reread the relative section of the book that provides a much more detailed analysis, and then use whatever solution is best for your soil.

Cation Exchange Capacity

The cation exchange capacity is largely a function of your soil texture, which you can't easily change. Clay soil will have higher values and will probably not be a problem. Low values indicate a sandy soil. The best way to increase the CEC is to add organic matter to the soil. Follow the recommendations below to increase OM levels.

Organic Matter

Nature increases organic matter over thousands of years. Although steps can be taken to speed it up, this is a slow process but one that needs to be started as soon as possible.

The best way to increase the organic matter in existing landscape beds is to use an organic mulch. Compost and manure are best for adding OM quickly, but they are not great at stopping weeds. A mulch of wood chips is better at controlling weeds, but it is a slow way to increase the OM level.

In new landscape beds or vegetable gardens, you can apply several inches of compost or manure and dig it into the soil. This disturbance of the soil is not good for soil health, but it's a compromise between damaging soil structure and adding the organic matter. It is best to do this once, as you prepare the bed. In following years, use mulch to continually increase the organic level.

pH

PH values that are too high or too low can be corrected and should be matched to the plants you will be growing. For most plants, a pH between 5.5 and 8 is be suitable, and it is better not to try to fine-tune the pH unless it is outside this range. Acid-loving plants need a pH below 6.

Use lime to raise the pH. The amount you use will depend on your buffer pH value; follow recommendations from your soil test. Use sulfur to lower a high pH, and apply it annually to maintain it.

Sodium Adsorption Ratio (SAR)

A high SAR value indicates a high level of sodium, which will prevent aggregation. If this is a problem it should be at the top of your priority list.

The first step is to figure out why sodium is high and then solve that problem. The cause is most likely a high saltwater table or irrigation with seawater.

The next step is to add gypsum to the soil, which will replace the sodium ions in clay with calcium ions. The testing lab will provide recommendations for the amount of gypsum to use.

Nutrient Levels

It is relatively quick to make these changes by following the advice of the testing lab. They will recommend fertilizers, but if you prefer the organic route, you will have to find organic amendments that provide the missing nutrients and apply them.

Keep in mind that most organic sources of nutrients are slow-acting. It can take several years before the amendment releases the nutrient in plant-useable forms. In cases such as rock dust, it might be 100 years before plants benefit.

High Levels of Sodium

In order to solve high sodium levels, you need to know why they are high. Do you have a high water table? Do you have high evaporation rates and the ground water is high in sodium? Was seawater used for irrigation?

If there is an identifiable reason for the high levels, try to eliminate the problem so that it does not occur again. Sodium is fairly easy to wash out of soil with irrigation, ideally with fresh water. Start a process of regular irrigation.

Aggregate Stability

Aggregate stability is improved using the same steps as for increasing the degree of aggregation.

Compaction

Compaction is difficult to fix in the short term. For really heavy compaction, some quick improvement can be achieved with rototilling or cultivating, along with the incorporation of organic matter. The best approach is to add organic matter and mulch. Then plant and let the plants and soil life improve the compaction over time.

Prevent further compaction by not working wet soil and staying off the soil as much as possible.

Degree of Aggregation

Aggregation is not something you can improve directly. It is something that is created by having healthy soil microbiology. Keep the microbes happy and aggregation will develop, and this is done by increasing OM, decreasing tillage, and improving compaction.

Earthworm Count

Earthworms feed on dead organic material and microbes. They collect this material from the surface of the soil and take it down into lower levels. Leaving the surface covered with plant refuse, adding organic matter to soil, using mulch to keep soil cool, eliminating tilling, and reducing compaction all have a positive effect on the growth of earthworms.

Exposed Soil

Exposed soil is a simple problem to fix—cover the soil. Leave more plant refuse on the soil. In established gardens, apply organic mulch such as wood chips or compost. In a vegetable garden, add mulch like straw as soon as your planting is complete and seeds have germinated. When fallow, keep the soil covered.

Hardpan

A hardpan prevents water from percolating through soil properly and stops plant roots from growing deep enough. A shallow hardpan should be a priority.

It is probably best to consult a specialist about the problem. It is essential to understand the severity of the hardpan and the cause of it, before you try to solve the issue.

Percolation

To some extent, the percolation rate is based on soil texture, which can't be easily changed, but percolation of fine soil can be improved by increasing aggregation and decreasing compaction, both of which produce larger pore spaces. This increases the amount of air in soil that can be replaced with water.

Plant Health

Poor plant growth can indicate a soil that is not very healthy, but more importantly this can show problems that are not addressed by other tests. If the standard chemical soil test doesn't indicate a problem, you should order an extended soil test or a plant tissue analysis to see if there is an issue with micronutrients.

Pesticide contamination can also cause poor growth or deformed plants, though this is a less likely cause.

Poor growth may also indicate poor cultural practices, such as watering incorrectly or using contaminated water. This usually shows itself on different kinds of plants and is not restricted to a particular species.

Smell

The smell test is a simple way to estimate the amount of air in soil. A lack of air leads to anaerobic conditions, which give off a sour or putrid smell. At the other extreme, soil with lots of microbe activity has a pleasant earthy smell.

Anything that improves the aeration of soil, and the growth of microbes, will improve the smell. This includes adding OM, reducing compaction, increasing aggregation, and keeping the soil covered with vegetation.

Salt Crust

Salt crust is usually caused by a high evaporation rate. The best solution is to increase the irrigation rate and/or reduce evaporation with mulch.

Soil Crusting

A surface crust indicates poor soil structure or a structure that is not very stable. In both cases, mulch will improve things. In the short term, it prevents rain from forming the crust; in the long term, it adds organic material to the soil, which will lead to better soil structure.

Soil Respiration

Soil respiration measures microbial activity and indirectly measures the number of microbes. A low value indicates a low number of microbes, which is usually the result of a low level of organic matter. Heavy compaction and low nutrient levels will also contribute to the problem.

Increase OM levels by using mulch, cover crops, and crop rotation. Reduce tillage.

Tighty Whitie Soil Test

The Tighty Whitie soil test measures microbe activity. Soil with a lot of microbes will decompose cotton fibers faster. A low microbiology can be improved by increasing OM, decreasing compaction, and keeping soil cool with mulch.

Action Plan

After reviewing the above solutions, select the items that will become your priority action items, and fill in the Action Plan table in Appendix B. Under the actions column, enter the tasks you will undertake, being as specific as possible. Then add a due date to each action to indicate when you will complete each task. This table becomes your marching orders for the next year.

Action Plan Follow-up

The plan for improving soil should be re-evaluated annually, preferable in early spring. Each year, complete another assessment, develop an action plan, and then carry out the tasks.

Example of an Action Plan.

Action Plan for Year: 2021   Date: February 2021

Site Description: Backyard vegetable garden

Problem	Actions	Due Date
Compacted soil	Create raised beds, without walls, that are 4 feet wide. Remove 4" of soil from the pathways and add it to the beds.	March 15
	Mulch pathways with straw.	March 15
Low OM level	Add 3" of compost on the raised beds, and dig it into the top 6" of soil, before planting.	March 15
	After planting, mulch with straw to keep the soil covered.	April and May
Low potassium level	Add 2 pounds potassium nitrate to the whole garden at the same time as the OM is added.	March 15

You will start seeing some improvement in your soil health during the first year if you had nutrient deficiencies, but most other improvements will be slower. Although it might take 10 years before you see really good quality of soil, there will be small improvements each year, and plants will benefit right away in each subsequent year.

Soil Health Assessment Form

Date: _____

Site description: _____

_____

_____

_____

Site history over past three years: _____

_____

_____

_____

Topsoil depth: _____

Texture (p. 15):   % sand =   % clay =

	Current Value	Rating	Previous value
Test Results Obtained from a Laboratory			
Cation Exchange Capacity (CEC) (1=20 and above; 2=between 5 and 20; 3=less than 5) [p. 44]			
Organic matter (% by weight) (1=4% or more; 2=2 to 4%; 3=less than 2%) [p. 71]			
pH (1=5.5 to 8.0; 3=below 5.5 or above 8) [p. 25]			
Sodium adsorption ratio (SAR) (1=less than 13; 3=more than 13)			
Calcium (1=adequate amount; 3=very low or very high)			
Iron (1=adequate amount; 3=very low or very high)			
Magnesium (1=adequate amount; 3=very low or very high)			
Phosphorus (1=adequate amount; 3=very low or very high)			
Potassium (1=adequate amount; 3=very low or very high)			
Sodium (1=low amount; 3=high amount)			
Sulfur (1=adequate amount; 3=very low or very high)			

	Current Value	Rating	Previous value
DIY Soil Tests			
Aggregate stability (1=remain intact for > 5 minutes; 2=disintegrate after 1–5 minutes; 3=disintegrated in < 1 minute) [p. 24]			
Compaction (1=wire penetrates at least 1 foot; 2= wire penetrates at least 8"; 3=less than 8" penetration) [p. 114]			
Degree of aggregation (1=good aggregation; 2=minimal aggregation; 3=no aggregation) [p. 23]			
Earthworm count (# per cubic foot) (1=10 or more; 2=3 to 10; 3=less than 3) [p. 69]			
Exposed soil (1= less than 25%; 2=25-75%; 3= 75% or more)			
Hardpan (1=no hardpan; 2=deep hardpan; 3=shallow hardpan at 6" or less) [p. 115]			
Percolation test (1=a value of 1–3" per hour; 2=more than 3" per hour; 3=less than 1" per hour) [p. 118]			
Plant health (1=plants grow well; 2=plants are small, light green; 3=poor growth)			
Smell (1=earthy, rich; 2=minimal odor; 3=sour, putrid)			
Salt crust (1=none; 2=minimal; 3=significant) [p. 110]			
Soil crusting (1=no crust, 2=some crust, 3=crust over the whole surface) [p. 111]			
Soil respiration (1=3.5 to 4; 2=2.5 to 3.5; 3=less than 2.5) [p. 113]			
Tighty Whitie soil test (1=only the elastic is left; 2=more than half is gone; 3=less than half is gone) [p. 112]			

Action Plan for the Year:

Date: _____

Site Description: _____

_____

_____

_____

Problem	Actions	Due Date

Index

About the Author

ROBERT PAVLIS, a Master Gardener with over 45 years of gardening experience, is owner and developer of Aspen Grove Gardens, a six-acre botanical garden featuring 3,000 varieties of plants. A popular and well-respected speaker and teacher, Robert has published articles in *Mother Earth News*, *Ontario Gardening* magazine, the Ontario Rock Garden Society website (Plant of the Month column), and local newspapers. He is also the author of the widely read blogs GardenMyths.com, which explodes common gardening myths, and GardenFundamentals.com, which provides gardening and garden design information. Robert also has a gardening YouTube channel called Garden Fundamentals.

Connect with Robert Pavlis

You can connect with me through social media by leaving comments in one of the following:

gardenmyths.com/
gardenfundamentals.com/
youtube.com/Gardenfundamentals1

The best way to reach me directly is through my Facebook Group, where I answer questions on a daily basis:
facebook.com/groups/GardenFundamentals/.

Also by the Author

Building Natural Ponds is the first step-by-step guide to designing and building natural ponds that use no pumps, filters, chemicals, or electricity and mimic native ponds in both aesthetics and functionality. Highly illustrated with how-to drawings and photographs.

For more information and ordering details,
visit BuildingNaturalPonds.com.

If you enjoyed this book, you may also like his other books, *Garden Myths Book 1* and *Garden Myths Book 2*. Each one examines over 120 horticultural urban legends.

Turning wisdom on its head, Robert Pavlis dives deep into traditional garden advice and debunks the myths and misconceptions that abound. He asks critical questions and uses science-based information to understand plants and their environment. Armed with the truth, Robert then turns this knowledge into easy-to-follow advice. Details about the books can be found at gardenmyths.com/garden -myths-book-1/.

They are available from Amazon and other online outlets.

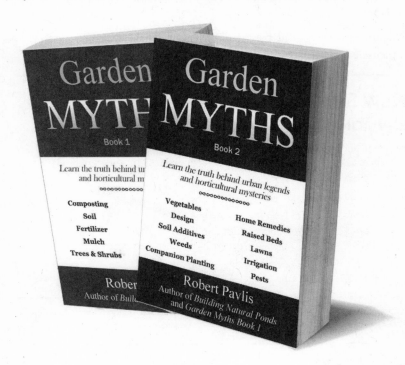

ABOUT NEW SOCIETY PUBLISHERS

New Society Publishers is an activist, solutions-oriented publisher focused on publishing books for a world of change. Our books offer tips, tools, and insights from leading experts in sustainable building, homesteading, climate change, environment, conscientious commerce, renewable energy, and more—positive solutions for troubled times.

We're proud to hold to the highest environmental and social standards of any publisher in North America. When you buy New Society books, you are part of the solution!

- We print all our books in North America, never overseas.

- All our books are printed on **100% post-consumer recycled paper**, processed chlorine-free, with low-VOC vegetable-based inks (since 2002).

- Our corporate structure is an innovative employee shareholder agreement, so we're one-third employee-owned (since 2015).

- We're carbon-neutral (since 2006).

- We're certified as a B Corporation (since 2016).

At New Society Publishers, we care deeply about *what* we publish—but also about *how* we do business.

Download our catalog at https://newsociety.com/Our-Catalog or for a printed copy please email info@newsocietypub.com or call 1-800-567-6772 ext 111.

New Society Publishers
ENVIRONMENTAL BENEFITS STATEMENT

By using 100% post-consumer recycled paper vs virgin paper stock, New Society Publishers saves the following resources:[1] (per every 5,000 copies printed)

23	Trees
2,050	Pounds of Solid Waste
2,255	Gallons of Water
2,941	Kilowatt Hours of Electricity
3,726	Pounds of Greenhouse Gases
16	Pounds of HAPs, VOCs, and AOX Combined
6	Cubic Yards of Landfill Space

[1] Environmental benefits are calculated based on research done by the Environmental Defense Fund and other members of the Paper Task Force who study the environmental impacts of the paper industry.

Certified
B Corporation

MIX
Paper from responsible sources
FSC www.fsc.org FSC® C016245

new society
PUBLISHERS
www.newsociety.com